Studies in Computational Intelligence 414

Editor-in-Chief

Prof. Janusz Kacprzyk
Systems Research Institute
Polish Academy of Sciences
ul. Newelska 6
01-447 Warsaw
Poland
E-mail: kacprzyk@ibspan.waw.pl

T0134999

For further volumes:
http://www.springer.com/series/7092

Alexander Gelbukh and Olga Kolesnikova

Semantic Analysis of Verbal Collocations with Lexical Functions

Authors
Alexander Gelbukh
CIC - Centro de Investigación
en Computación
IPN - Instituto Politécnico Nacional
Av. Juan Dios Bátiz s/n esq. Av. Mendizábal
Col. Nueva Industrial Vallejo
Mexico

Olga Kolesnikova
CIC - Centro de Investigación
en Computación
IPN - Instituto Politécnico Nacional
Av. Juan Dios Bátiz s/n esq. Av. Mendizábal
Col. Nueva Industrial Vallejo
Mexico

ISSN 1860-949X
ISBN 978-3-642-43633-8
DOI 10.1007/978-3-642-28771-8
Springer Heidelberg New York Dordrecht London

e-ISSN 1860-9503
ISBN 978-3-642-28771-8 (eBook)

Printed on acid-free paper

Springer is part of Springer Science+Business Media (www.springer.com)

Abstract

Lexical function is a concept that formalizes semantic and syntactic relations between lexical units. Relations between words are a vital part of any natural language system. Meaning of an individual word largely depends on various relations connecting it to other words in context. Collocational relation is a type of institutionalized lexical relations which holds between the base and its partner in a collocation (examples of collocations: *give a lecture, make a decision, lend support* where the bases are *lecture, decision, support* and the partners, termed collocates, are *give, make, lend*). Collocations are opposed to free word combination where both words are used in their typical meaning (for example, *give a book, make a dress, lend money*). Knowledge of collocation is important for natural language processing because collocation comprises the restrictions on how words can be used together.

There are many methods to extract collocations automatically but their result is a plain list of collocations. Such lists are more valuable if collocations are tagged with semantic and grammatical information. The formalism of lexical functions is a means of representing such information. If collocations are annotated with lexical functions in a computer readable dictionary, it will allow their precise semantic analysis in texts and their effective use in natural language applications including parsers, high quality machine translation, periphrasis system and computer-aided learning of lexica. In order to achieve these objectives, we need to extract collocations from corpora and annotate them with lexical functions automatically.

To this end, we use sets of hypernyms to represent the lexical meaning of verbal collocations and evaluate many supervised machine learning techniques on predicting lexical functions of unseen collocations. to train algorithms, we created a dictionary of lexical functions containing more than 900 disambiguated and annotated examples. The dictionary is a part of this book. The obtained results show that machine learning is feasible to achieve the task of automatic detection of lexical functions.

About the Authors

Alexander Gelbukh received a PhD degree (Computer Science) from the All-Russian Institute of Scientific and Technical Information (VINITI) in 1995. Since 1997 he is a Research Professor and Head of the Natural Language and Text Processing Laboratory of the Center for Computing Research of the National Polytechnic Institute (IPN), Mexico, since 1998 he is a National Researcher of Mexico, currently with excellence level 2, and since 2000 he is an member of the Mexican Academy of Sciences. He is autor of more than 400 publications, and editor of more than 50 books and special issues of journals, in the areas of computational linguistics and artificial intelligence. See more details on www.Gelbukh.com.

Olga Kolesnikova received her PhD degree (Computer Science, with honors) from the National Polytechnic Institute (IPN), Mexico, in 2011. She was graduated from, and currently collaborates with, the Natural Language and Text Processing Laboratory of the Center for Computing Research of the IPN. She has received (jointly with A. Gelbukh) the Best Paper Award from the Mexican International Conference on Artificial Intelligence in 2011, and the Best PhD Thesis in Artificial Intelligence Award from the Mexican Society of Artificial Intelligence for the thesis "Automatic extraction of lexical functions" (advisor A. Gelbukh), in 2011. She is author of various journal papers and book chapters in the areas of linguistics and computational linguistics.

Contents

1 Introduction 1
   1.1 Text and Computer 2
   1.2 Lexical Function 4
   1.3 Lexical Resources..8
      1.3.1 Resources of English Collocations 11
      1.3.2 Problems 12
      1.3.3 Solution 13
   1.4 Glossary...13

2 Lexical Functions and Their Applications..19
   2.1 Theoretical Framework..19
   2.2 Lexical Functions and Relations between Words 20
      2.2.1 Institutionalized Lexical Relations...20
      2.2.2 Collocational Relations..21
   2.3 Definition and Examples of Lexical Functions.....................................24
      2.3.1 More about Lexical Function $Oper_1$...28
      2.3.2 Lexical Functions Caus and Liqu 29
      2.3.3 Lexical Functions for Semantic Derivates 29
      2.3.4 Lexical Functions for Semantic Conversives..............................31
      2.3.5 Several Other Lexical Functions...31
   2.4 Lexical Functions and Collocations..32
      2.4.1 Syntagmatic Lexical Functions..32
      2.4.2 Syntagmatic Lexical Functions as a Collocation Typology..........33
   2.5 Lexical Functions in Applications 34
      2.5.1 Word Sense Disambiguation..34
         2.5.1.1 Syntactic Ambiguity 35
         2.5.1.2 Lexical Ambiguity 36
      2.5.2 Computer-Assisted Language Learning.......................................36
         2.5.2.1 Game "Lexical Function" 36
      2.5.3 Machine Translation 37
         2.5.3.1 Implementation Example 37
      2.5.4 Text Paraphrasis..38
         2.5.4.1 Applications of the Lexical Functions and
               Paraphrastic Rules..39

3 Identification of Lexical Functions...41
   3.1 Lexical Functions in Dictionaries 41
      3.1.1 Explanatory Combinatorial Dictionary 42

3.1.2 Combinatorial Dictionaries for Machine Translation 43
3.1.3 French Combinatorial Dictionaries..45
3.1.4 Dictionary of Spanish Collocations 46
3.2 Automatic Identification of Lexical Functions 46
3.2.1 Research in (Wanner, 2004) and (Wanner *et al.*, 2006)................47
3.2.2 Research in (Alonso Ramos *et al.*, 2008).....................................49
3.2.3 Three Hypothesis Stated by Wanner *et al.* (2006) 49
3.3 Automatic Detection of Semantic Relations...50

4 Meaning Representation **51**
4.1 Linguistic Meaning and its Representation...51
4.2 Hypernym-Based Meaning Representation 53
4.2.1 Hypernyms and Hyponyms..53
4.2.2 WordNet as a Source of Hypernyms...54
4.2.3 An Example of Hypernym-Based Meaning Representation 55
4.3 Semantic Annotations as Meaning Representation...................................57

5 Analysis of Verbal Collocations with Lexical Functions **61**
5.1 Verbal Lexical Functions...61
5.1.1 Basic Lexical Functions in Verbal Collocations...........................61
5.1.2 Lexical Functions Chosen for our Experiments............................65
5.2 Supervised Machine Learning Algorithms 66
5.2.1 WEKA Data Mining Toolkit 67
5.2.2 Stratified 10-Fold Cross-Validation Technique 67
5.3 Representing Data for Machine Learning Techniques...............................67
5.3.1 Training (Data) Sets..68
5.4 Classification Procedure 70
5.5 Algorithm Performance Evaluation 73
5.5.1 Baseline...74
5.5.2 Three Best Machine Learning Algorithms for Each Lexical
Function 75
5.5.3 Algorithm Performance on the Training Set.................................77
5.5.4 Three Hypothesis in (Wanner *et al.*, 2006) and our Results 77
5.5.5 Best Machine Learning Algorithms...79
5.5.6 LF Identification k-Class Classification Problem 81

6 Linguistic Interpretation..**85**
6.1 Computer Experiments in Linguistics 85
6.2 Linguistic Statement: Our Hypothesis...86
6.2.1 Collocational Isomorphism..86
6.2.2 Collocational Isomorphism Represented as Lexical Functions 88
6.3 Testing the Linguistic Statement 89
6.4 Statistical and Semantic View on Collocation...90
6.5 Conclusions 91

7 **Dictionary of Spanish Verbal Lexical Functions** **93**
    7.1 Compilation 93
    7.2 Description..95

References **97**

Appendix 1: Experimental Results..**101**

Appendix 2: Dictionary of Spanish Verbal Lexical Functions.......................**117**

Chapter 1
Introduction

Human language is a unique, perfect and beautiful instrument of communication and cognition. It is unique because no other known living being possesses such a rich variety of phonetic, lexical and syntactical means of expressing themselves and in fact do not have a need to do so. It is perfect because it gives us a truly limitless capacity of shaping our thoughts, conveying our emotions, inventing fresh ideas, describing our experience, putting in a concise and consistent form our knowledge of the universe. The beauty of language is revealed in picturesque, vivid and elegant works of poetry and literature.

Communication today has become much wider and diverse due to modern web technologies. Now we can not imagine how the previous generations were able to organize their life and resolve political, economic, social problems without computers, cell phones, Internet. One of the greatest advantages of modern technologies is that now we are all connected, and this connection is not limited by city or country boundaries, but works on a world wide basis, uniting us all in the global community so that we can share what we have learnt or experienced with all the humankind and in turn have access to all the intellectual and cultural treasures the humankind has accumulated.

The benefit of possessing an immense amount of information is accompanied by a big challenge of making this information not only accessible but also useful in the sense that we can operate on this huge knowledge collection in various ways: search through it, make summaries, translate texts of our interest into other natural languages, extract evidence, principles, ideas and opinions, answer questions, confirm or refute our hypothesis and guesses and fulfill other tasks related to intelligent information processing. How to accomplish these tasks automatically and efficiently? The solution will inevitably deal with developing computer applications that "understand" the meaning of information units, or, in other words, are capable of making a complex and multi-faceted semantic analysis of information. Our book describes one of the approaches to automatic semantic analysis of information expressed in human language. This approach is based on the linguistic formalism of lexical function and machine learning techniques and has been developed for analyzing a particular type of language structures called

A. Gelbukh, O. Kolesnikova: Semantic Analysis of Verbal Collocations, SCI 414, pp. 1–18.
springerlink.com   © Springer-Verlag Berlin Heidelberg 2013

verbal collocations, for example, *to give a smile*, *to lend support*, *to take a walk*. **Verbal collocations** are those combinations of the verb with other words (mainly, prepositions, nouns, and adverbs) whose meaning can not be predicted on the basis of the meanings of their elements. In three verbal collocations that we have just given, all three verbs (*to give*, *to lend*, *to take*) have the meaning 'to perform'[1] but this meaning is lexicalized by different verbs depending on the respective noun. A *smile* chooses *to give* for the meaning 'to perform (a smile)', *support* prefers *to lend* to express the same meaning, and *a walk* selects *to take* for the same purpose. Such unpredictive choices are typical of collocations.

In this chapter, we will first introduce the reader to the area of research that looks for effective strategies and develops robust methods for intelligent processing of human language. Next, we will present in an informal way the concept of lexical function as a tool for semantic analysis of language and then we will speak of lexical resources necessary for automatic processing of language.

This book is written for both linguists and computer scientists working in the field of artificial intelligence as well as to anyone interested in intelligent text processing. Since this branch of research is interdisciplinary and its terminology is borrowed from such distinct domains as language studies and computational technologies, readers may find it helpful to consult unfamiliar terms in Glossary which concludes this Introduction.

1.1 Text and Computer

The majority of information encountered on the World Wide Web is expressed in human language or, as it is conventional nowadays to say, **natural language.** The term *natural language* is opposed to *artificial language* created in order to operate and control machines, most commonly computers. Speaking of computers, an artificial language (for example, C++) has a favorable position compared to a natural language (for example, English) because artificial words are not ambiguous, that is, they have only one meaning. Thus it is guaranteed that computer will understand any command perfectly and fulfill it.

On the other hand, words of natural language may have many meanings, and sometimes very different. What is the meaning of a *tongue*? Is it "a movable mass of muscular tissue attached to the floor of the mouth in most vertebrates" or "a flap of leather on a shoe, either for decoration or under the laces or buckles to protect the instep", or "a language, dialect or idiom"? (Meaning definitions are cited the *Collins English Dictionary* at http://www.collinsdictionary.com/dictionary/english). Taken as such, the word *tongue* is ambiguous, that is, it has more than one meaning, and outside context there is no way to decide which one of its meanings is to be applied.

The process of assigning a word its appropriate meaning within a given context is called **ambiguity resolution** or **disambiguation**. It is one of the biggest and

---

[1] At this point and further on, single inverted commas are used to indicate the meaning of the enclosed words.

most difficult problems of **natural language processing (NLP)**, computer-based manipulation of information expressed in natural language which we know as **text**.

The issue of disambiguation is very important for many tasks of automatic text processing, and we will mention and briefly describe some of them.

Perhaps, the most known and certainly most used natural language application is **information retrieval** systems. Imagine our Internet connections fail and suddenly the most popular web search engines like Google or Yahoo! Search can not work, how upset and nervous we will become! Information retrieval has truly become a part of our everyday life and the most popular way to find answers to our questions, resolve doubts, learn new things and just have a good time.

Information extraction is a kind of information retrieval and its goal is to find in text documents and withdraw specific pieces of data like names, relations, terminology, images, audio, video on particular topics. Knowledge acquisition in its broad sense is the process of acquiring knowledge possessed by humans for the purpose of creating expert systems.

Question answering may also be considered a type of information retrieval, but instead of typing queries and getting a huge list of links to   voluminous texts, one asks a question and is expected to receive an exact and complete answer. A question may be *How many Summer Olympic Games has the United States hosted*? A well-designed question answering system must output *Four*.

Automatic knowledge acquisition aims at developing applications that deduct rules as a form of knowledge representation from sentences in natural language. Such applications are incorporated into complex knowledge-based systems that model human cognitive functions (finding associations, pattern recognition, generalization, specification, etc.) and simulate problem solving and decision making processes. The mission of expert systems is to assist in resolving challenging and demanding tasks in medicine and biotechnology, finances, industry, transport, resources management and other fields.

Machine translation can be called the oldest natural language application. It reflects the dream and the hope of multi-language global community to overcome language barriers. Machine translation is a classical example of language processing and may be thought as the "barometer" of success in solving the disambiguation problem explained previously.

Clearly, the disambiguation problem is a matter of a great scientific interest within the area of artificial intelligence, but to a common user, it turns into a very practical concern. In some cases, when we expect a computer application to give us a quality assistance in translating a text or finding a good answer to a question, our expectations are disappointed by erroneous and confusing outputs.

For example, the phrase *The child could not solve this problem, but he deserves some credit for trying to do it* is rendered into Spanish by Google translator as *El niño no puede resolver este problema pero se merece algún crédito para tratar de hacerlo*. Ignoring minor discrepancies of this translation, we can notice that the error is in the wrong literal translation of *credit* as *crédito*. *Credit*, like in the above example of *tongue*, is an ambiguous word, and in the context of the given utterance means 'approval', not 'money', and thus should be translated as *mérito*, not *crédito*.

Another example is *We should take the responsibility for this act* translated in Spanish as *Debemos tomar la responsabilidad de este acto*. Although *tomar la responsabilidad* (lit. *to take the responsibility*) seems to be lexically and grammatically correct as well as adequately understood, to a native Spanish speaker it sounds awkward and mechanically fabricated as if it were a molded copy of the corresponding English expression.

What is wrong with *tomar la responsabilidad*? The matter is that the noun *responsabilidad* prefers the verb *asumir* to *tomar* in order to express the meaning 'to take (the responsibility)' so that *asumir la responsabilidad* is not only correct but naturally sounded Spanish equivalent of the latter English phrase. Such expressions in which one word chooses another one to convey a particular meaning in an unmotivated, unpredicted way are called **collocations** or **restricted lexical co-occurrence**. The reader will come across a little bit different way of explaining and illustrating the notion of collocation in Section 1.2.

Indeed, why do we say *to make a decision*, but *to give a walk*, not **to make a walk*? Why do we make such a lexical choice though *to make* and *to give* in these combinations seem to have the same meaning which can be put as 'do, realize, carry out (something)'?

Well, while the reason for this unmotivated and semantically unpredictable linguistic phenomenon can not yet be fully explained, we have to deal with it somehow in order to avoid incorrect interpretation or generation of collocations in computer applications and rule out expressions like **to make support* instead of *to lend support* or **to carry out a laugh* instead of *to give a laugh* or **to give a survey* instead of *to carry out a survey*.

An efficient tool for semantic processing of collocations is lexical function. This formalism allows to represent the meaning of collocations in an exact, concise and original manner, to disclose and give explicit formalization of the association between a particular meaning and a specific syntactic structure, to introduce a proper order in what seemed to be a chaotic heap of astonishingly diverse word combinations and to produce a harmonious and simple classification of collocations. Now we are going to introduce the reader to this powerful concept.

1.2 Lexical Function[2]

In this section, we introduce the concept of lexical functions with illustrations followed by informal definition and explanation. A formal definition and more details will be given in Section 2.3.

It does not surprise us that *a bank* can be a financial institution as well as a piece of land. It is quite often that one word is used with different meanings. But

---

[2] Written with I.A.Bolshakov.

sometimes the opposite happens: we choose different words to express the same idea. For example, *to give a smile* means *to smile*, and *to lend support* means *to support*. These two combinations convey the same idea: *to smile* is *to "perform"*, or *"do"* *a smile*, and *to support* is *to "do" support*, so that both verb-noun combinations share the same semantics: to *do* what is denoted by the noun. Likewise we find that *to acquire popularity* and *to sink into despair* both mean 'to *begin to experience* the <noun>', and *to plant a garden* and *to write a letter* mean 'to *create* the <noun>'. Such semantic and syntactic patterns or classes are called **lexical functions**. They are generalizations of semantic and syntactic properties of large groups of word combinations, the majority of which are collocations. Thus, lexical function is an explicit and clear representation of selectional preferences of words.

Let us consider a multilingual example to illustrate the concept of lexical function. Observe the following three groups of synonymous phrases given in different languages:

1. Eng. **strong** *tea*
   Ger. **starker** ('powerful') *Tee*
   Sp. té **cargado** ('loaded')
   Fr. thé **fort** ('forceful')
   Rus. **krepkiy** ('firm') *chay*
   Pol. *herbata* **mocna** ('firm')

2. Eng. **ask** *a question*
   Sp. **hacer** ('make') *una pregunta*
   Fr.   **poser** ('put') *une question*
   Rus. **zadat'** ('give') *vopros*

3. Eng. **let out** *a cry*
   Sp. **dar** ('give') *un grito*
   Fr. **pousser** ('push') *un cri*
   Rus.**ispustit'** ('let out') *krik*

Here, the restriction on lexical co-occurrences is impossibility of certain word combinations in one language while the same combination is absolutely normal in another language. For example, though in Spanish the combinations like **té duro*, **dar una pregunta*, **dejar salir un grito* are impossible, we can't account for this by absurdity of their meaning, since in another language (Russian, to say) they sound perfectly. Similarly, a perfect Spanish phrase *té cargado* sounds absurdly in literal English translation **loaded tea*.

In Example 1, there is nothing in the meaning of the well-prepared tea that can explain why one should qualify it as *strong* in English, *powerful* in German, *loaded* in Spanish, *forceful* in French and *firm* in Russian and Polish. Similarly, in Examples 2 and 3, the selection of the appropriate verb to be used together with a given noun cannot be done basing on the meaning of the latter: there is nothing in the meaning of *question* to explain why in English you *ask* it, in Spanish you *make* it, in French you *put* it, and in Russian you *give* it, but never *ask*, *make*, or *put*.

As we see, this type of restrictions is a property of the corresponding languages and thus is to be studied in the frame of linguistics. In fact, this type of restrictions on lexical co-occurrences is one of the main objectives of lexicographic description. To reflect this type of restrictions, the concept of lexical functions is used.

The notion of lexical functions is closely related to the notion of collocations. To give some informal insight into the latter concept, we will explain in with the following examples.

The meaning of word combination such as *to give a book* or *to lend money* can be obtained by mechanically combining the meaning of the two constituting words: if *to give* is 'to hand over', *a book* is 'a pack of pages' then *to give a book* is 'to hand over a pack of pages'. However, the meaning of such word combinations as *to give a lecture* or *to lend support* is not obtained in this way: *to give a lecture* is not 'to hand it over'. Such word pairs are called **collocations**. Collocations are difficult for a computer system to analyze or generate because their meaning cannot be derived automatically from the meaning of their constituents. However, this difficulty could be resolved, if each collocation is assigned its respective lexical function, since the latter represents the collocational meaning.

Now using the concept of collocation, we can define **lexical function** as a semantic and structural relation between two constituents of a collocation: the **base**, a word used in its typical meaning, and the **collocate**, a word semantically dependent on the base. In collocation *to make a mistake*, the base is *a mistake*, and the collocate is *to make*, in *to establish a relation* the base is *a relation*, and the collocate is *to establish*.

The word **function** in the term **lexical function** characterizes well the type of relation between the base and the collocate as opposed to the elements of free word combinations. Mathematically, a function is a mapping from one object to another such that one object necessarily corresponds to another object. The cohesion between collocational elements is strong enough so that the base chooses a particular word, and not another one, to express certain meaning. So we *make mistakes*, but do not *do* them, or *a lecture is delivered*, but not *made* or *done*. Semantically independent word, or the base, is called the **keyword** of the lexical function. The keyword is mapped unto the collocate which is called the **value** of the lexical function. Thus a collocation can be represented in a mathematical form:

$$\textbf{LexicalFunction}(keyword) = value$$

Lexical functions capture semantic content of collocations and structural relations between the base and the words dependent on it. The collocational semantics is represented by the meaning of a corresponding lexical function. Every lexical function is given a name in the form of an abbreviated Latin word which semantically is closest to the meaning of a given lexical function. For example, the lexical function **Oper** (from Lat. *operare*, do) has the abstract meaning 'do, perform, carry out something'. This meaning is present in the collocations *to* *deliver a lecture, to make a mistake, to lend support, to give an order,* therefore, they can be represented with the lexical function **Oper**: **Oper**(*lecture*) = *to deliver*, **Oper**(*mistake*) = *to make*, **Oper**(*support*) = *to lend*, **Oper**(*order*) = *to give*. Other collocations like *to create a difficulty, to hold elections, to sow suspicions, to give surprise* can be formalized by the lexical function **Caus** (from Lat. *causare*, cause) which has the meaning 'to cause something to come into existence'. Therefore, the collocations mentioned above can be represented as **Caus**(*difficulty*) = *to create*, **Caus**(*elections*) = *to hold*, **Caus**(*suspicions*) = *to sow*, **Caus**(*surprise*) = *to give*.

Besides the meaning, lexical functions represent structural relations between the base of the collocation and the words semantically dependent on it. For example, *lecture* implies that a person *delivers* it to another person(s), that is, the interpretation of *lecture* includes the agent (a person who delivers a lecture) and the recipient (a person or, typically, people who listen to a lecture). When the collocation *to deliver a lecture* is used in utterances, the agent is normally the subject. In the notation of lexical functions this fact is encoded by the subscript 1 of the corresponding lexical function, in this case, **Oper**: **Oper**$_1$(*lecture*) = *to deliver*. The same syntactic structure is seen in the collocation lend support: **Oper**$_1$(*support*) = *to lend*. But the collocation *to receive support* is represented as **Oper**$_2$(*support*) = *to receive*, because the subject in the utterances with *to receive support* is the recipient identified by subscript 2.

Up to now we have considered collocations where the value of the lexical function, i.e., the collocate, includes one semantic element: *to do* in **Oper** and *to make* in **Caus**. But often the meaning of the collocate contains more that one semantic element. For example, in the collocation *to lead* (somebody) *to the opinion*, the collocate *to lead* contain both semantic elements: 'cause' and 'do'. This phenomenon can be represented by complex lexical functions which are combinations of simple lexical functions considered previously in this section.

Thus, *to lead* (somebody) *to the opinion* can be represented as **CausOper**$_1$(*opinion*) = *to lead*. In this case, **Caus** does not have any index or subsript, meaning that the object that caused the opinion is neither the agent, nor the recipient of opinion, and **Oper** has the subscript 1 meaning that the person who was lead to the opinion, is the agent of the opinion. The collocation *to put* (somebody) *under* (somebody's) *control* is represented as **CausOper**$_2$(*control*) = *to put*, where 2 specified that the person put under control is the recipient of control, not its agent. Lexical functions like **Oper**, **Caus** which include one semantic element are called simple elementary lexical functions, and complex lexical functions encode more than one semantic element.

More than 70 elementary lexical functions have been identified. Each lexical function represents a group of word combinations possessing the same semantic content and syntactic structure. If we construct a database of word combinations annotated with semantic and syntactic data in the form of lexical functions, natural language processing systems will have more knowledge at their disposal to fulfill such an urgent and challenging task as word sense disambiguation.

1.3 Lexical Resources

Lexical resources are widely used in natural language processing and their role is difficult to overestimate. Lexical resources vary significantly in language coverage and linguistic information they include, and have many forms: word lists, dictionaries, thesauri, ontologies, glossaries, concordances.

Somehow the word *dictionary* has turned into a general term to name almost any lexical resource whose "skeleton" is an alphabetically ordered word list, where each word is followed by some information. It is a typical dictionary organization by entries where a word is the title, and the body of an entry contains information which may include pronunciation, variants of orthography, definition(s), equivalents in another language(s), examples of usage, typical structures, indication of domain, style, etymology, dialect and other features. The choice of a specific dictionary depends on the application, and it is not rare that language engineers compile special dictionaries intended to be used in their particular application.

The most known and commonly used for NLP tasks is WordNet (Miller, 1998). Originally, this lexical database was developed for English in Princeton University. It is a freely available lexical resource, and the latest version WordNet 3.1 is accessible online at http://wordnet.princeton.edu (Princeton University, 2010).

The main organizational unit of WordNet is not a word, but a set of words with very similar meanings called **synonyms**. Therefore, an entry is entitled by a synonym set termed **synset**. An example of a synset is {*car#1*, *auto#1*, *automobile#1*, *machine#6*, *motorcar#1*}. Since a word (a string of letters) may have many meanings or senses, each word in a synset is followed by its sense number. The meaning of the whole synset is called **concept**.

The advantage is such an organization is that every word in the synset has only one meaning—the meaning of the concept—described by a definition called **gloss** (thus it is not ambiguous!) and one or more examples of usage of word(s) in the synset. So the synset *car, auto, automobile, machine, motorcar* is followed by its gloss: a motor vehicle with four wheels; usually propelled by an internal combustion engine, and a short phrase illustrating its usage in context: *"he needs a car to get to work"*.

It seems that there is nothing special about this rather simple structure. But what has given WordNet such an immense significance in NLP is a system of relations between its overall 117,000 synsets. A relation is in fact a labelled pointer from one synset to another where the label specifies the type of relation. The following major types of relations are indicated in WordNet:

- **hypernymy** or **IS-A relation** indicates that a word (*bicycle*) belongs to a particular class of objects (*wheeled vehicle*). Thus, a *bicycle* IS A *wheeled vehicle*. A more general word is called **hypernym**, and a more specific one, **hyponym**;
- **meronymy** or **PART-WHOLE relation** connects a word denoting an object with another one denoting a part of this object (*wheel – bicycle*);

- **troponymy** is a relation between verbs, it characterizes an action by some particular manner of performing it (walk-step);
- **antonymy** is a relation between adjectives with opposite meaning (*clear*#1 – *unclear*#2, *clear*#10 – *ill-defined*#1).

As we have seen, a word in WordNet is disambiguated by its synset and characterized by a set of relations with other synsets (words). This information is of extreme value for practically any NLP application. That is why for some widely used in NLP programming languages like Java, Perl, Ruby, special modules have been developed for quering WordNet.

The importance of WordNet for research is shown by a simple fact that the query "WordNet" gives 41,700 hits in Google Scholar with the option "in articles and patents". In contrast, the query "Collins English Dictionary" generates 4,010 hits, and "Oxford Advances Learner's Dictionary" produces 2,800 hits (the search was made on December 21, 2011).

The practical value of WordNet has been appreciated by computational linguistis all around the world and the movement of creating WordNets for other languages was started by launching the EuroWordNet (Vossen, 1998) which initially was designed for Dutch, Italian, Spanish, German, French, Czech, and Estonian. Later, wordnets were developed for Swedish, Norway, Danish, Greek, Portuguese, Basque, Catalan, Romanian, Lithuanian, Russian, Bulgarian, Slovenic. Currently, the development of wordnets continues for even more languages including Arabic, Chinese, Korean, Japanese, Urdu, Sanskrit, Hebrew, among others.

Wordnets are used in so many diverse computational tasks that it is hard to list them all, bearing in mind that while this book is in the process of its creation, new tasks are continuously appearing. To our current knowledge, Wordnets are successfully applied in document and text retrieval, text classification and categorization, sentiment detection and opinion mining, image retrieval and annotation, open-domain question answering (much more complicated that answering question for a specific and narrow domain!), semantic role labeling, textual entailment, word sense disambiguation and many other tasks.

We have mentioned in the previous section that disambiguation is at the heart of natural language processing. Collocations are especially challenging units of language because of their unpredictable and semantically unmotivated nature. Now, how are the collocations represented in WordNet?

To answer this question, we first need to clarify the difference between collocations from **idioms** or **fixed expression** (*to kick the bucket, to rise from the ashes*). The latter have non-compositional nature and the meaning of the whole phrase can not be analyzed by the meaning of its components. Idioms are represented in WordNet as **words-with-spaces** (as if it were not a phrase but a signle word), for example:

(v) spill the beans#1, let the cat out of the bag#1, talk#5, tattle#2, blab#1, peach#1, babble#4, sing#5, babble out#1, blab out#1 (divulge confidential information or secrets) *"Be careful--his secretary talks"*

However, collocations do not possess such a strong idiomatic or non-compositional nature; the meaning of a collocation (*to make a mistake*) is closer to the meaning of its elements than in an idiom (*to spill the beans*). Still, the choice of the collocate (*to make*, why not *to do*?) is not motivated to such a degree as in free word combinations (*to make a dress/cake/toy*), that is why it is not a trivial task to analyze collocations correctly, and that is the reason why learners of English as a second language do not sound so natural (we do not mean their pronunciation, but lexical choices they make).

In spite of their importance, collocations are not represented in WordNet in a consistent way, there is no single well-defined approach to identify, classify and describe collocations. Sometimes they appear in the examples section of an entry (to make a mistake):

> (n) mistake#1, error#1, fault#1 (a wrong action attributable to bad judgment or ignorance or inattention) *"he made a bad mistake"; "she was quick to point out my errors"; "I could understand his English in spite of his grammatical faults"*

The same collocation is placed in the entry of the verb *to make*, thus creating redundancy:

> (v) make#24 (carry out or commit) *"make a mistake"; "commit a faux-pas"*

Collocations occasionally appear in glosses (*to deliver a lecture*):

> (v) lecture#1, talk#6 (deliver a lecture or talk) *"She will talk at Rutgers next week"; "Did you ever lecture at Harvard?"*

Such collocations as *to lend support, to hold elections, to establish a relation, to sink into despair, to give a laugh* are not present in WordNet. The absence of collocational data makes it necessary to infer selectional preferences from corresponding entries with the help of some additional information, perhaps, knowledge of semantic field, synonym/hypernym relations, structural patterns found in corpora. Consider, for example, that we wish to generate the collocation *to give a laugh* using the following entries:

> (n) laugh#1, laughter#1 (the sound of laughing)
> (n) laugh#2 (a facial expression characteristic of a person laughing) *"his face wrinkled in a silent laugh of derision"*

> (n) joke#1, gag#1, laugh#3, jest#1, jape#1 (a humorous anecdote or remark intended to provoke laughter) *"he told a very funny joke"; "he knows a million gags"; "thanks for the laugh"; "he laughed unpleasantly at his own jest"; "even a schoolboy's jape is supposed to have some ascertainable point"*

> (v) give#7, throw#5 (convey or communicate; of a smile, a look, a physical gesture) *"Throw a glance"; "She gave me a dirty look"*

Certainly, at first we have to decide what sense of *laugh* is present in *to give a laugh* which does not seem a very easy matter. Then, we are lucky enough to find *smile* in the gloss of *give#7*. The last step is to deduce that if *to give a smile* is a correct and naturally sounded combination, them *to give a laugh* is one also.

To deal with such difficulties, we need dictionaries or lexical databases of collocations. Now we are going to consider some of them.

1.3.1 Resources of English Collocations

The biggest and most well-known dictionary of English collocations is **The Oxford Collocations Dictionary for Students of English** (McIntosh *et al.*, 2009). It includes 250,000 word combinations and 75,000 examples. Each entry in this dictionary is entitled with a word like in general-purpose dictionaries, but the body of an entry consists of collocations typical for this word grouped according to their structure, for example, adjective + noun, verb + noun, verb + adjective, preposition + adjective, etc.

Syntactic information plays a big role in text analysis, but as to the meaning of collocations and their semantic characteristics, this dictionary can not be of help to language engineers. Certainly, this dictionary was not created for NLP purposes but only for learning English as testifies its title, but it is a pity that such an excellent repository of collocations, in the form that it exists, can not serve well enough for computational tasks.

Macmillan Collocations Dictionary for Learners of English (Rundell, 2010) includes collocates for 4,500 headwords. One of the advantages and truly a big "plus" of this dictionary is that collocations are divided into semantic sets, and each set is provided with a brief definition. Thus, this lexicon is not only a source of syntactic information but also of meaning.

Concerning its usage for NLP, the dictionary has certain limitations. First, it is written with a particular objective to teach English upper-intermediate and advanced English learners how to produce natural and idiomatic academic and professional writing, so its domain is bounded to, first, written English, then, within the realm of written English, to academic and professional writing. Hence the choice of headwords and their collocates is guided by this purpose. Secondly, it is not designed as a machine-readable dictionary, thus it needs serious adaptations and modifications before it can be usable in NLP applications.

Another method of learning collocational properties of words is to use **online corpus query systems** that provide the option of retrieving collocations of a given word. One such system, Sketch Engine (www.sketchengine.co.uk), gives open access to ACL Anthology Reference Corpus (ARC), British Academic Spoken English Corpus (BASE), British Academic Written English Corpus (BAWE), and Brown Corpus. This software has a function of constructing so-called word sketches, i.e. combinations of a particular headword with other words in corpus grouped by their grammatical relations (noun + noun, verb + adverb, etc.). Many of these combinations are collocations.

Another example of online corpus query system is the tool created by Mark Davies (http://view.byu.edu). His system works on four large English corpora: Corpus of Contemporary American English (COCA), Corpus of Historical American English (COHA), TIME Magazine of American English, and British

National Corpus. This program includes the option of retrieving collocation for a given headword.

To our opinion, these online systems have three basic limitations. First, the corpora are not big enough, i.e., does not contain the number of words big enough to make conclusions concerning typical or representative collocations of the English language as a whole. Secondly, they are not big enough with respect to their domains which are not so wide to consider them as open-domain corpora. Thirdly, what is extracted by such systems are in fact collocation candidates, and not all of them are true collocations. The list of collocation candidates should be examined by a human expert in order to get rid of noise, i.e., word combinations incorrectly assigned the status of collocations but formally correct according to the requirements of a specific collocations recognition technique. Besides, the percentage of collocations in the output of a retrieval system (precision) depends on particular metrics used for collocation extraction. A comprehensive survey of metrics used in collocation extraction methods can be found in (Pecina, 2005).

1.3.2 Problems

In the previous section we mentioned some problems of lexical resources regarding their treatment of collocations, and here we will list them briefly:

- insufficient semantic annotation,
- lack of consistency in collocations representation,
- limit in size (not broad enough),
- narrow domain (not representative enough of a language).

Another problem of both dictionaries and corpora is that they become "old" very quickly. Compilation of dictionaries and quality corpora requires a lot of human effort and time, so these databases are not capable of capturing newly appearing words. Also with time, after the creation process has been completed and these products become accessible to open public, some of words and expressions included in them might already have become obsolete, or gone out of common use.

Therefore, there exists a great need of creating **electronic collocation dictionaries** "on the fly", automatically, using all textual data at hand, i.e., all electronic documents and all that is published on the World Wide Web. Besides, collocations should be annotated with semantic and syntactic information, to enable their precise and correct analysis.

The lack of lexical resources and its negative inhibiting effect on NLP development has been increasingly emphasized in literature. Here are some quotes from research articles on language engineering: "...the lack of large domain-independent lexica is a major bottleneck for the development of more knowledge-rich NLP applications" (Burchardt et al., 2006); "NLP suffers from a lack of lexical resources" (Navarro et al., 2009); "The lack of large-scale, freely available and durable lexical resources, and the consequences for NLP, is widely acknowledged but the attempts to cope with usual bottlenecks preventing their development often result in dead-ends" (Sajous *et al.*, 2010).

Though computerized lexicography has achieved a significant progress over last years, compilation of high quality dictionaries still requires a lot of manual work. We mentioned previously, that in such a multi-faceted area as computational linguistics, it is difficult sometimes to find an adequate lexical resource (and of the language you need) appropriate for a specific research task or application. One way to solve this problem is to develop computational procedures which can adjust existing resources to the demands of a researcher. However, this is not always effective. Certainly, the best solution (but certainly not the easiest!) of this problem is to compile a new lexical resource.

1.3.3 Solution

Lexical function is a powerful theoretical instrument for collocations systematization and analysis. This formalism captures not only syntactic properties, but, what are even more important, semantic properties of collocations. Lexical funcitons make it possible to put in order an immence quantity of collocations according to their meaning and structure. Moreover, if lexical functions are automatically identified in a sufficiently precise manner, language technologies can be developed which permit to recognise new collocations and thus keep lexical databases up-to-date.

In this book, we will present a thorough exposition of the concept of lexical functions, give more detail of such functions that cover verbal collocations, examine machine learning techniques on the task of lexical functions detection, and present a dictionary of lexical functions for Spanish verbal collocations.

1.4 Glossary

Adjective – in English, a word which meets certain inflectional tests (e.g. takes comparative **–er** or superlative **–est**) and distributional tests (e.g. precedes a noun, follows a degree word); typically functions to qualify or quantify a noun, e.g. *slender, happy, original* (Brinton and Brinton, 2010:393).

Adverb – in English, a word that is sometimes inflected for degree, may be modified by a degree word, and functions to modify all parts of speech except nouns, e.g., *fast, smoothly, happily, consequently* (Brinton and Brinton, 2010:393).

Agent – a thematic role that may be borne by a noun phrase in a sentence. It is the agent that is construed as undertaking the action of the sentence (Weisler and Milekic, 2000:305).

Artificial intelligence – the study of systems that act in a way that to any observer would appear to be intelligent. It involves using methods based on the intelligent behavior of humans and other animals to solve complex problems (Coppin, 2004:xv).

Complement – the same as **Direct object**.

Clause – a grammatical unit that includes, at minimum, a predicate and an explicit or implied subject, and expresses a proposition. The following example sentence contains two clauses: *It is cold, although the sun is shining*. The main clause is *it is cold* and the subordinate clause is *although the sun is shining* (Loos, 1997).

Determiner – a position in grammatical structure that is filled by various items that specify something about how the audience is to identify the referent of a particular noun. In English, determiners include the articles (a, the, absence of a/the), demonstratives (this, that, these, those), all possessors, some question words, pronouns, etc. (Payne, 2006:326).

Direct object – a grammatical relation that exhibits a combination of certain independent syntactic properties, such as the following: the usual grammatical characteristics of the patient of typically transitive verbs, a particular case marking, a particular clause position, the conditioning of an agreement affix on the verb, the capability of becoming the clause subject in passivization, the capability of reflexivization.The identification of the direct object relation may be further confirmed by finding significant overlap with similar direct object relations previously established in other languages. This may be done by analyzing correspondence between translation equivalents (Loos, 1997).

Grammar – the internalized, unconscious system of conventions and corres-pondences that allows people to communicate with one another. It consists of everything speakers must know in order to speak their language. Linguistic analysis consists in making this implicit system explicit, usually in the form of written rules and concise statements (Payne, 2006:332).

Grammatical category – a semantic distinction often expressed by inflectional morphology or periphrasis (function words), e.g., number (singular vs. plural), tense (present vs. past) (Brinton and Brinton, 2010:401).

Grammatical relations – grammatically instantiated relations between words in phrases or clauses. Some typical grammatical relations are genitive, subject, object, ergative, absolutive, and others (Payne, 2006:331).

Grammatical rule – a regular pattern in the grammar that determines how a conceptual category is expressed structurally (for example "add the suffix –**ed** to form the past tense of a verb") (Payne, 2006:331).

Idiosyncratic – unpatterned, random. For example, the plural of *child* in English is idiosyncratic, *children*, in that there are no other nouns in the modern language that form their plurals in precisely this way(Payne, 2006:332).

Lexeme – the minimal unit of language which has a semantic interpretation and embodies a distinct cultural concept, it is made up of one or more form-meaning composites called lexical units (Loos, 1997).

Lexical category – a syntactic category for elements that are part of the lexicon of a language. These elements are at the word level. Lexical category is also known as part of speech, word class, grammatical category, grammatical class. Lexical categories may be defined in terms of core notions or "prototypes". Given forms may or may not fit neatly in one of the categories. The category membership of a form can vary according to how that form is used in discourse (Loos, 1997).

Lexical form – an abstract unit representing a set of wordforms differing only in inflection and not in core meaning (Loos, 1997).

Lexical relation – a culturally recognized pattern of association that exists between lexical units in a language (Loos, 1997).

Lexical semantic relations – semantic relations between concepts (Chiu *et al.*, 2007).

Lexical unit – a form-meaning composite that represents a lexical form and single meaning of a lexeme (Loos, 1997).

Lexis (lexical level of language, lexical means of language) – vocabulary of language. The study of lexis is the study of the vocabulary of languages in all its aspects: words and their meanings, how words relate to one another, how they may combine with one another, and the relationships between vocabulary and other areas of the description of languages (the phonology, morphology and syntax) (Malmkjær, 2002:339).

Machine learning – a field of study which seeks to answer the question "How can we build computer systems that automatically improve with experience, and what are the fundamental laws that govern all learning processes?" (Mitchell, 2006)

Meaning – a notion in semantics classically defined as having two components: reference, anything in the referential realm denoted by a word or expression, and sense, the system of paradigmatic and syntagmatic relationships between a lexical unit and other lexical units in a language (Loos, 1997).

Morphology (morphological level of language, morphological means of language) – morphological The next level of structure is the morpheme, the smallest unit of meaning in language. Rules of morphology focus on how words (and parts of words) are structured and describe all facets of word formation, such as how prefixes and suffixes are added (Meyer, 2009:7-8).

Noun – a word denoting a person, an animal, a concrete o abstract thing, e.g., *child* (person), *goose* (animal), *table* (concrete thing), *happiness* (abstract thing).

Object (also see **Direct object**) – A core grammatical relation, defined in English by the following properties: (1) Position immediately following the verb in pragmatically neutral, transitive clauses, (2) when pronominalized, non-subject pronouns are used, and (3) absence of a preceding preposition (Payne, 2006:337).

Patient – the semantic role held prototypically by entities that undergo a visible, concrete change in state, e.g., *the cake* referred to in *Alice ate the cake* (Payne, 2006:338).

Paradigmatic lexical relation – a culturally determined pattern of association between lexical units that share one or more core semantic components, belong to the same lexical category, fill the same syntactic position in a syntactic construction, and have the same semantic function (Loos, 1997). "In general, paradigmatic relations subsume all contrast and substitution relations that may hold between lexical units in specific contexts. For example, the lexemes school and student are paradigmatically related in such pairs of phrases as to go to school and to be a student, and so also are the lexemes young and tall in young student and tall student. A paradigmatic relation, in general, does not automatically imply a semantic relation" (Wanner, 1996).

Phonetics (phonetic level of language, phonetic means of language) – This level focuses on the smallest unit of structure in language, the phoneme (distinctive speech sounds; that is, they create meaningful differences in words). Linguistic rules at this level describe how sounds are pronounced in various contexts (Meyer, 2009: 7, 196).

Pragmatics – the study of how context affects and is affected by linguistic communication (Payne, 2006:340).

Predicate – the portion of a clause, excluding the subject, that expresses something about the subject, e.g., in the sentence *The book is on the table* the predicate is *is on the table* (Loos, 1997).

Preposition – an indeclinable word which governs an object, e.g. *in* the morning, *after* the movie, *of* the book (Brinton and Brinton, 2010:409).

Prototype of a category – the member or set of members of a category that best represents the category as a whole. Not everything fits perfectly in a category. Categories are defined by an intersection of properties that make up their members. Members that have all the properties are the prototype members. Those that contain some, but not all, of the properties are less prototypical (Loos, 1997).

Recipient – the semantic role held prototypically by entities that receive some item, e.g., the message-world participant referred to by the phrase *the carpenter* in *The walrus gave a sandwich to the carpenter* (Payne, 2006:341).

Relation – that feature or attribute of things which is involved in considering them in comparison or contrast with each other; the particular way in which one thing is thought of in connexion with another, any connexion, correspondence, or association, which can be conceived as naturally existing between things (Little *et al.*, 1959).

Subject – a grammatical relation that exhibits certain independent syntactic properties, such as the following: the grammatical characteristics of the agent of typically transitive verbs, the grammatical characteristics of the single argument of intransitive verbs, a particular case marking or clause position, the conditioning of an agreement affix on the verb, the capability of being obligatorily or optionally deleted in certain grammatical constructions (such as the following clauses: adverbial, complement, coordinate), the conditioning of same subject markers and different subject markers in switch-reference systems, the capability of coreference with reflexive pronouns. The identification of the subject relation may be further confirmed by finding significant overlap with similar subject relations previously established in other languages. This may be done by analyzing correspondence between translation equivalents (Loos, 1997).

Syntax (syntactic level of language, syntactic means of language, syntactic structure) – is concerned with how words arrange themselves into constructions (Malmkjær, 2002:354).

Semantics – is the study of linguistic meaning, and is the area of linguistics which is closest to the philosophy of language. The main difference between the linguist's and the philosopher's way of dealing with the question of meaning is that the linguist tends to concentrate on the way in which meaning operates in language, while the philosopher is more interested in the nature of meaning itself – in particular, in the relationship between the linguistic and the non-linguistic (Malmkjær, 2002:455).

Semantic component – a potentially contrastive part of the meaning of a lexical unit. Contrastive semantic component distinguishes one lexical unit from another, for example, "male" is the contrastive semantic component distinguishing man from woman, and boy from girl; shared semantic component occurs in each member of a group of lexical units, for example, "human" is a shared component for man, woman, boy, and girl (Loos, 1997).

Semantic (thematic) role – aspects of interpretation assigned to noun phrases and other parts of linguistic constituents that indicate, for example, who is doing what to whom. The assignment of thematic roles is also an important feature of syntactic analysis. Agent is an example of a thematic role — the agent of the sentence is who/what undertakes action. *John* bears the thematic role of agent in the sentence *John sat on Bill* (Weisler and Milekic, 2000:328).

Syntagmatic lexical relation – a culturally determined pattern of association between pairs of lexical units (A1-B1, A2-B2, A3-B3...) where the two members of each pairs (A1 and B1) have compatible semantic components, are in a fixed syntactic and semantic relationship to each other, and are typically associated with each other, and corresponding members of each pair (A1, A2, A3...) belong to the same lexical category, fill the same syntactic position in a syntactic construction, and have the same semantic function (Loos, 1997).

Syntactic head – the element of a phrase that determines (or "projects") the syntactic properties of the whole phrase, e.g., *cat* in *That ridiculous big orange cat that always sits on my porch*. In this case the syntactic head also happens to be the semantic head. However, in some cases the two can be distinct. For example, prepositions are syntactic heads because they determine the syntactic behavior of their phrases, even though the nominal component expresses most of the meaning (Payne, 2006:331).

Syntactic categories – a cover term for all the types of units that figure into a syntactic structure. Syntactic categories include lexical categories, phrasal categories, and in earlier versions of Generative grammar, the category S, or Sentence (Payne, 2006:345).

Syntactic pattern – a way of expressing conceptual categories that involves the addition of one or more free morphemes, or an adjustment in the order of free morphemes. These are also called analytic or periphrastic patterns (Payne, 2006:345).

Syntactic structure – the linear order, constituency, and hierarchical relationships that hold among linguistic units in an utterance (Payne, 2006:345).

Syntagmatic relations – (or co-occurence relations) hold between lexical units that can appear together, i.e. that co-occur, in the same phrase or clause (Wanner, 1996).

Token – each running word in the text. Thus a text of length a hundred words contains a hundred tokens (Sinclair *et al.*, 2004).

Verb – a word denoting an action, a state or a process, e.g., action: *to buy, to find, to come*; state: *to understand, to have, to enjoy*; process: *to eat, to read, to walk*.

Word class – traditionally called "parts of speech." They are grammatically distinct classes of lexical items, such as nouns, verbs, adjectives, adverbs, etc. (Payne, 2006:348).

Word form – the phonological or orthographic sound or appearance of a word that can be used to describe or identify something, the inflected forms of a word can be represented by a stem and a list of inflections to be attached (*WordNet 3.0*).

Chapter 2
Lexical Functions and Their Applications[*]

2.1 Theoretical Framework

As a concept, lexical function (LF) was introduced in the frame of the Meaning-Text Theory (MTT) presented in (Mel'čuk, 1974, 1996) in order to describe lexical restrictions and preferences of words in choosing their "companions" when expressing certain meanings in text. Here we will give a brief account of the fundamental concepts and statements of MTT as the context of LFs. Actually, the formalism of lexical functions has been one of those parts of MTT which attracted most attention of specialists in general linguistics and, in particular, computational linguistics. A lot of research began in the area of natural language processing with the purpose to develop various approaches and techniques of employing LFs in such applications as word sense disambiguation, computer-assisted language learning, machine translation, text paraphrases, among others.

Meaning-Text Theory (MTT) was created by I.A. Mel'čuk in the 1960s in Moscow, Russia. Nowadays, the ideas that gave rise to MTT are being developed by Moscow semantic school (Apresjan, 2008). MTT was proposed as a universal theory applicable to any natural language; so far it has been elaborated on the basis of Russian, English and French linguistic data.

MTT views natural language as a system of rules which enables its speakers to transfer meaning into text (speaking, or text construction) and text into meaning (understanding, or text interpretation). However, for research purposes, the priority is given to the meaning-text transfer, since it is believed that the process of text interpretation can be explained by patterns we use to generate speech. MTT sets up a multilevel language model stating that to express meaning, humans do not produce text immediately and directly, but the meaning-text transfer is realized in a series of transformations fulfilled consecutively on various levels.

Thus, starting from the level of meaning, or semantic representation, we first execute some operations to express the intended meaning on the level of deep syntax, then we go to the surface syntactic level, afterwards proceeding to the deep morphological level, then to the surface morphological level, and we finally arrive to the phonological level where text can be spoken and heard (oral text or

---

[*] Written with I.A.Bolshakov.

A. Gelbukh, O. Kolesnikova: Semantic Analysis of Verbal Collocations, SCI 414, pp. 19–40.
springerlink.com   © Springer-Verlag Berlin Heidelberg 2013

speech). Another option is a written text which is actually speech represented by means of an orthography system created to facilitate human communication. Each transformational level possesses its own "alphabet", or units, and rules of arranging units together as well as rules of transfer from this level to the next one in the series. So at each level we obtain a particular text representation – e.g., deep syntactic representation, surface morphological representation, etc.

Semantic representation is an interconnected system of semantic elements, or network; syntactic representations are described by dependency trees; morphological and phonological representations are linear.

The most significant aspects of MTT are its syntactic theory, theory of lexical functions and the semantic component – explanatory combinatorial dictionary. The syntactic part is most fully presented in (Mel'čuk 1993–2000), lexical functions are explained further in this chapter, and the best achievement in lexicography is explanatory combinatorial dictionaries for Russian and French (Mel'čuk and Zholkovskij, 1984; Mel'čuk *et al.*, 1984) described in Section 3.1.

2.2 Lexical Functions and Relations between Words

One of the purposes of the lexical function theory is its application to classification of collocations. Therefore, we successively explain the concepts of institutionalized lexical relations, collocational relations, lexical functions and syntagmatic lexical functions.

2.2.1 Institutionalized Lexical Relations

Wanner (2004) states, that lexical function is a concept which can be used to systematically describe "institutionalized" lexical relations. We will consider the notion of institutionalized lexical relations first and show its relevance to collocation.

Wanner clarifies that "**a lexical relation is institutionalized** if it holds between two lexical units L_1 and L_2 and has the following characteristics: if L_1 is chosen to express a particular meaning M, its choice is predetermined by the relation of M to L_2 to such an extent that in case M and L_2 is given, the choice of L_1 is a language-specific automatism."

Institutionalized lexical relations can be of two types: paradigmatic and syntagmatic. **Paradigmatic relations** are those between a lexical unit and all the other lexical units within a language system (as between synonyms, hypernyms and hyponyms, etc.) and **syntagmatic relations** are those between a lexical unit and other lexical units that surround it within a text. These are some examples of words between which there exist institutionalized lexical relations: *feeling – emotion, move – movement, snow – snowflake, acceptance – general, argument – reasonable, make – bed, advice – accept*, etc. In the first three pairs of words we observe paradigmatic institutionalized lexical relations and the syntagmatic ones, in the rest of the examples.

In the above definition the term **lexical uni**t is used to define a form-meaning composite that represents a lexical form and single meaning of a lexeme (Loos, 1997). The phrase "institutionalized lexical relation" does not have the meaning of the relation between lexical units which build a cliché as in (Lewis, 1994). Speaking of multi-word entities as a whole, Lewis divides them into two groups: institutionalized expressions and collocations. He defines collocations as made up of habitually co-occurring individual words. Then he adds that collocations tell more about the content of what a language user expresses as opposed to institutionalized expressions which tell more about what the language user is doing, e.g. agreeing, greeting, inviting, asking, etc.

Institutionalized lexical relations in (Wanner, 2004) possess the quality of association present between habitually co-occurring words, or collocations, if we consider the syntagmatic type of institutionalized lexical relations. Indeed, as it can be seen from definitions of collocation which we will discuss in the next section of this chapter, that the relations between collocation components, i.e. between the base and the collocate, are characterized by high probability of occurrence of the collocate given the base and by arbitrariness of the collocate choice. The latter is emphasized by (Lewis, 1997) who insists that collocations do not result from logic but are decided only by linguist convention.

We can notice that the concept of institutionalized lexical relations is wider than the concept of relations between collocation elements. As it was mentioned in the beginning, institutionalized lexical relations can be paradigmatic and syntagmatic. Paradigmatic relations connect lexical units along the vertical axis of a language system while syntagmatic relations connect words positioned along the horizontal axis of the language; they tie together lexical units within a linear sequence of oral utterance or a written text as, for example, they link the base and its collocate in a collocation. So the relations within a collocation are institutionalized syntagmatic lexical relations which can be also termed collocational relations.

2.2.2 Collocational Relations

Wanner (1996) gives the following definition of collocational relation. "A **collocational relation** holds between two lexemes L_1 and L_2 if the choice of L_1 for the expression of a given meaning is contingent on L_2, to which this meaning is applied. Thus, between the following pairs of lexical units collocational relations hold: *to do*: *a favor*, *to make*: *a mistake*, *close*: *shave*, *narrow*: *escape*, *at*: *a university*, *in*: *a hospital*." The term **lexeme** used in the above definition is the minimal unit of language which has a semantic interpretation and embodies a distinct cultural concept, it is made up of one or more form-meaning composites called lexical units (Loos, 1997).

As it is seen from the definition, a collocational relation holds between components of non-free word combinations, i.e. such combinations whose semantics is not fully compositional and has to be partially or entirely derived from the phrase as a whole. Non-free combinations are opposed to free word combinations where syntagmatic relations hold between words in a phrase with purely compositional semantics.

Examples of free word combinations are: *a black cable, a different number, to put a chair [in the corner], to write a story, to run quickly, to decide to do something*. Examples of non-free word combinations are: *a black box, as different again, to put into practice, to write home about, to run the risk of [being fired], to decide on [a hat]*. The distinction between free and non-free combinations is a general distinction usually made in linguistic research with respect to syntagmatic relations.

Collocational relations can be classified according to lexical, structural and semantic criteria. The most fine-grained taxonomy of collocations based on semantic and structural principle was given by Mel'čuk (1996). This taxonomy uses the concept of lexical function which we are going to consider in the next section.

However, before speaking of lexical functions, we will list some of the existing definitions of collocations and give a brief characteristic of them. These definitions are elaborated within different linguistic trends and, when put together, help to understand better such a multi-sided linguistic phenomenon as collocations. The list is structured as follows. First, a definition is given with its corresponding reference, then, it is indicated what criterion for a word combination to be considered as collocation is implied by this definition, and then some comments are given.

- Collocations of a given word are statements of the habitual or customary places of that word (**Firth, 1957**). **Lexical criterion** (a word is used in a fixed position with respect to another element of collocation) and **statistical criterion** (frequency of word co-occurrence). Firth was the first to introduce the term 'collocation' from Latin *collocatio* which means 'bringing together, grouping'. He believes that speakers make 'typical' common lexical choices in collocational combinations. Collocation is a concept in Firth's theory of meaning: "Meaning by collocation is an abstraction at the syntagmatic level and is not directly concerned with the conceptual or idea approach to the meaning of words. One of the meanings of *night* is its collocability with *dark*, and of *dark*, of course, collocation with *night*."

- Collocation is the syntagmatic association of lexical items, quantifiable, textually, as the probability that there will occur, at *n* removes (a distance of *n* lexical items) from an item *x*, the items *a, b, c ...* (**Halliday, 1961**). **Lexical criterion** (a word is used in a fixed position with respect to another element of collocation) and **statistical criterion** (high co-occurrence frequency). If a lexical item is used in the text, then it's collocate has the highest probability of occurrence at some distance from the lexical item. Collocations cut across grammar boundaries: e.g., the phrases *he argued strongly* and *the strength of his argument* are grammatical transformations of the initial collocation *strong argument*.

- Collocations are binary word-combinations; they consist of words with limited combinatorial capacity, they are semi-finished products of language, affine combinations of striking habitualness. In a collocation one partner determines, another is determined. In other words, collocations have a basis and a co-occurring collocate (**Hausmann, 1984**). **Lexical** criterion (the lexical choice of the collocate depends on the basis). All word combinations are classified into two basic groups, i.e., fixed and non-fixed combinations, with further subdivisions, and in this classification, collocations belong to the category of non-fixed affine combinations. Internal structure of collocation is emphasized, i.e., collocation components have functions of a basis and a collocate, and the basis (not the speaker) 'decides' what the collocate will be.

- Collocation is a group of words that occurs repeatedly, i.e., recurs, in a language. Recurrent phrases can be divided into grammatical collocations and lexical collocations. **Grammatical collocations** consist of a dominant element and a preposition or a grammatical construction: *fond of*, (*we reached*) *an agreement that...* **Lexical collocations** do not have a dominant word; their components are "equal": *to come to an agreement, affect deeply, weak tea* (**Benson et al., 1986**). **Functional criterion** (collocations are classified according to functions of collocational elements) and **statistical criterion** (high co-occurrence frequency). This understanding of collocation is broad, and collocations are classified according to their compositional structure.

- Collocations should be defined not just as 'recurrent word combinations', [but as] 'ARBITRARY recurrent word combinations' (**Benson, 1990**). **Lexical criterion** (arbitrariness and recurrency). 'Arbitrary' as opposed to 'regular' means that collocations are not predictable and cannot be translated word by word.

- Collocation is "that linguistic phenomenon whereby a given vocabulary item prefers the company of another item rather than its 'synonyms' because of constraints which are not on the level of syntax or conceptual meaning but on that of usage" (**Van Roey, 1990**). **Statistical criterion** (high co-occurrence frequency in corpora). Van Roey summarizes the statistical view stated by Halliday in terms of expression or 'usage'. A collocate can thus simply be seen as any word which co-occurs within an arbitrary determined distance or *span* of a central word or *node* at the frequency level at which the researcher can say that the co-occurrence is not accidental. This approach is also textual in that it relies solely on the ability of the computer program to analyze large amounts of computer-readable texts.

- Collocations are associations of two or more lexemes (or roots) recognized in and defined by their occurrence in a specific range of grammatical constructions (**Cowie, 1994**). **Structural criterion** (collocations are distinguished by patterns). Collocations are classified into types according to their grammatical patterns.

- Collocations are composite units which are placed in a Howarth's lexical continuum model on a sliding scale of meaning and form from relatively unrestricted (collocations) to highly fixed (idioms). Restricted colloca-tions are fully institutionalised phrases, memorized as wholes and used as conventional form-meaning pairings (**Howarth, 1996**). **Syntactic criterion** (commutability: the extent to which the elements in the expression can be replaced or moved (*to make/reach/take decision* vs. *to shrug one's shoulders*). **Semantic criterion** (motivation: the extent to which the semantic origin of the expression is identifiable, e.g., *to move the goalposts* = to change conditions for success **vs**. *to shoot the breeze* = to chatter, which is an opaque idiom). Classification includes four types of expressions with no reference to frequency of occurrence:

- o   free collocations (*to blow a trumpet* = to play a trumpet),
- o   restrictive collocations (*to blow a fuse* = to destroy a fuse/to get angry),
- o   figurative idioms (*to blow your own trumpet* = to sell oneself excessively),
- o   pure idioms (*to blow the gaff* = to reveal a concealed truth).

The problem with this classification is that it is difficult to determine what is meant by 'syntactically fixed', 'unmotivated' or 'opaque'. This is seen in the previous ambiguous example of *to blow a fuse*.

- •   Collocation is the co-occurrence of two items in a text within a specified environment. Significant collocation is a regular collocation between two items, such that they co-occur more often than their respective frequencies. Casual collocations are "non-significant" collocations. (**Sinclair** *et al.*, **2004**). **Lexical criterion** (recurrency of co-occurrence) and **statistical criterion** (high co-occurrence frequency). The degree of significance for an association between items is determined by such statistic tests as Fischer's Exact Test or Poisson Test.

- •   Collocation is a combination of two lexical items in which the semantics of one of the lexical items (the base) is autonomous from the combination it appears in, and where the other lexical item (the collocate) adds semantic features to the semantics of the base (**Mel'čuk, 1998**). Gledhill (2000) explains that for Mel'čuk, a collocation is a **semantic function** operating between two or more words in which one of the words keeps its 'normal' meaning. **Semantic criterion** (the meaning of a collocation is not inferred from the meaning of the base combined with the meaning of the collocate). Semantics of a collocation is not just the sum of the meaning of the base and the meaning of the collocate, but rather the meaning of the base plus some additional meaning that is included in the meaning of the base. '...the concept of collocation is independent of grammatical categories: the relationship which holds between the verb *argue* and the adverb *strongly* is the same as that holding between the noun *argument* and the adjective *strong*' (Fontenelle, 1994).

2.3 Definition and Examples of Lexical Functions

It was mentioned previously, that the concept of lexical function was first introduced in (Mel'čuk, 1974) within the frame of the Meaning-Text linguistic model. It is a way to organize collocational data with the original purpose of text generation. Here we will supply the definition of lexical function from (Mel'čuk, 1996).

"The term **function** is used in the mathematical sense: $f(X) = Y$. ...Formally, a Lexical Function f is a function that associates with a given lexical expression L, which is the argument, or keyword, of f, a set $\{L_i\}$ of lexical expressions – the value of f – that express, contingent on L, a specific meaning associated with f:

$$f(L) = \{L_i\}.$$

Substantively, a Lexical Function is, roughly speaking, a special meaning (or semantico-syntactic role) such that its expression is not independent (in contrast to all "normal" meanings), but depends on the lexical unit to which this meaning applies. The core idea of Lexical Functions is thus lexically bound lexical expression of some meanings."

Now we will illustrate the above given definition with examples and expose it in the manner similar to the way LFs were presented in Section 1.2, so that the reader might get more familiarized with the concept. Here we will introduce the lexical function **Oper₁**, the same one shown in Section 1.2, only now the explanation is wider, it discloses more aspects of the concept and gives it another perspective.

To begin with, let us consider two sentences where the verb *to give* is used:

(1) *He gave you a gift.*
(2) *He gave you a smile.*

In sentence (1), the meaning of the phrase *to give a gift* is composed of the meaning of *to give* and the meaning of *a gift*. That is, the meaning of this phrase is a simple sum of the meaning of its components. This is true for free word combinations to which *to give a gift* surely belongs. If we observe the phrase *to give a smile* in (2), we notice that its meaning can not be represented as a sum of the meanings of its components. The noun *a smile* is used in its common or most frequent meaning.

However, this is not the case of the verb *to give*. It adds such a semantic component to the meaning of *to give a smile* that makes the whole phrase convey the idea of *to smile* or "to carry out" *a smile*. It can also be mentioned that in order to express the meaning *to carry out* with *a smile*, only the verb *to give* is chosen, and in fact this is the only choice because no other verb would be used in such a case by a native English speaker. This restrictive word co-occurence is characteristic of collocations to which *to give a smile* belongs. The same phenomenon is observed in the following sentences:

(1) *Take a pen.*
(2) *Take a walk.*

In the first sentence, we see a free word combination, and in the second sentence, a collocation *to take a walk* which has the meaning "to carry out" *a walk*. Now, we have two collocations with the same semantic component "to carry out": *to give a smile* and *to take a walk*. Designating the semantic element **to carry out** by **Oper** (abbreviated Latin verb *operare,* to do) and using mathematical notation, we write the following:

Oper(*a smile*) = *to give*,
Oper(*a walk*) = *to take*.

Oper is the name of a lexical function, its value for the argument, or keyword, *a smile* is *to give*, for the argument *a walk*, *to take*. We can say that the generalized meaning or gloss of the lexical function **Oper** is **to carry out**. Here are other examples of **Oper**:

Oper(*a lecture*) = *to deliver*,
Oper(*an announcement*) = *to make*.

Oper values make up collocations with their respective keywords: *to deliver a lecture, to make an announcement*, where the keyword is the base and the LF value is the collocate.

Thus, lexical function **Oper** denotes the collocational relation with the gloss 'to carry out', 'to perform', 'to experience'. Other glosses, and therefore, lexical functions can be distinguished among collocational relations. Consider some LFs in Table 1.

Table 1 Examples of lexical functions

LF	Name description	Gloss	Keyword	Value	Collocation
Fact	Lat, *factum*, fact	to accomplish itself	*dream*	*come true*	*the dream came true*
Real	Lat. *realis*, real	to fulfill the requirement contained in the argument	*invitation*	*accept*	*to accept an invitation*
Caus	Lat. *causare*, to cause	to cause to exist	*association*	*found*	*to found an association*
Magn	Lat. *magnus*, big	intense, intensely, very (intensifier)	*temperature*	*high*	*high temperature*

Latin abbreviations are used for the names of lexical functions. The names are accompanied by numeric subscripts. They signify how the LF argument and the LF semantic structure are projected onto the syntactic structure of the LF value. For example, used as a subscript for **Oper,** 1 means that the Agent (the first participant) in the situation denoted by the argument is the verb's subject and the argument itself is its object. For example:

Oper$_1$(*a demand*) = *to present* (*to present a demand*).

More examples of **Oper**$_1$ are given in Table 2. Recall that the gloss of **Oper**$_1$ is **to carry out**. As a subscript for the same LF, 2 means that the verb's subject is the Undergoer (the second participant) in the situation denoted by the argument. For example:

Oper$_2$(*guilt*) = *to have* (*to have guilt*).

The gloss of **Oper$_2$** is **to undergo**. Zero (0) as a subscript for the lexical function **Func** (Latin *functionare*, to function) shows that the LF argument is the subject of an intransitive verbal value. For example:

Func$_0$(*wind*) = *to blow* (*the wind blows*).

Another way to explain the system of syntactic annotations is to say that some lexical functions and rules that use them rely on the notion of so-called ***actants*** of a word. This notion roughly corresponds to a *valency*: an actant is a complement of the word, describing necessary participants of the situation referred to by the word. For example, the situation referred to by the verb *to sell* involves the following roles: the seller, the buyer, the goods, and the price. Subsequently, they allow for the corresponding complements in the sentence: *The director did not approve the fact that the new computers had been sold by the institute to a French company for one thousand dollars*. Thus, this verb has four actants, or valencies.

The actants are referred to by numbers. The subject, or active agent (one who *is doing*, e.g., the seller) is the 1st actant, the object (that *upon what* it is done, e.g., the goods) is the 2nd actant, and the other actants are numbered according to their importance in the situation; their order is just a convention.

A lexical function is defined not for any lexeme. First, the arguments of a specific function often must belong to some set of lexemes, characterizing by a common component of their meanings, by their part of speech, or by some another property; the function is just not defined outside of this set. In most cases the domain set of a function is the same in different languages. Second, even if a word does belong to the domain set of the function, just by accident in a specific language there might not be the appropriate word, while in another language such a word might exist.

In other words, lexical functions are formalism for description of combinatorial abilities of individual lexemes. This description is topical on deep language levels, i.e., the deep syntactic and semantic ones.

A simpler version of the LF definition in (Mel'čuk, 1996) states that a lexical function **F** associates with its argument lexeme *L* another lexeme, or a set of quasi-synonymous lexemes, *L'* that expresses some standard abstract meaning and can play a specific role when used in the text in a syntactic relation with *L*.

Now we will consider examples of another lexical function, **Magn** (see its meaning and another example in Table 1). **Magn** specifies for a noun *N* an adjective, or a word combination of adjective type, which expresses the meaning of great intensity or magnitude of the main quality of *N*. Informally we can say that **Magn** (*L*) answers the question: *How to say "very" about L?* Thus, we can rewrite the phrases of the example given in Section 1.2 as

Eng. **Magn** (*tea*) = *strong*
Ger. **Magn** (*Tee*) = *starker*
Sp. **Magn** (*té*) = *cargado*
Fr. **Magn** (*thé*) = *fort*
Rus. **Magn** (*chay*) = *krepkiy*
Pol. **Magn** (*herbata*) = *mocna*

Here are several additional examples: **Magn** (*knowledge*) = *profound*, **Magn** (*appetite*)= *wolfish*, **Magn** (*thirst*) = *raging*. **Magn** is applicable not only to nouns but to many verbs, adjectives, and adverbs as well. Its argument should then have some main feature that can be qualified in grades, and correspondingly, **Magn** expresses the idea of the great degree of this feature, i.e., the meaning of 'very' or 'intensely'.

Notice that in any natural language there is some "standard", or the most usual, such adverb for adjectives and adverbs: *very* in English, *muy* in Spanish, *très* in French, *ochen'* in Russian. On the other hand, a non-native speaker should use these values very cautiously, since a lot of adjectives and verbs require quite different words in this meaning. That is why the true usage of **Magn**, as well as of other LFs, is so important for deep language competence.

About 70 so-called ***standard elementary*** LFs, such as **Magn** or **Oper**$_1$, were introduced and well elaborated in the Meaning-Text theory. A multiplicity of non-elementary functions can be built as combinations of the elementary ones as functions of other functions. The entire set of functions, both elementary and complex, allows for an exhaustive and highly systematic description of almost all language-dependent restrictions on lexical co-occurrences in natural languages.

2.3.1 More about Lexical Function Oper₁

Table 2 shows some examples of the function **Oper**$_1$. It is easy to see that the values of **Oper**$_1$ *do not have* any autonomous meaning for any of its arguments in both languages. Instead, they are merely tools to incorporate the substantive argument into the syntactic structure. Their role in the text is similar to that of auxiliary words; that is why they are termed **light** or **support verbs**.

Table 2 Examples of Oper$_1$

Argument	Oper₁	Glued verb
attention	*pay, focus, devote*	–
aid	*render, give, offer, provide, come to*	*aid*
help	*give, offer, provide*	*help*
lessons	*teach*	*teach*
concert	*give*	–
confidence	*have*	*trust*
cooperation	*give*	*cooperate*
pain	*feel, have*	*suffer*
cry	*let out*	*cry out*
fruits	*yield, bear*	*fruit*
necessity	*be under*	*need*
question	*ask*	*ask*
victory	*win, gain, achieve*	*conquer*

The proof of the absence of any meaning of the values of **Oper**$_1$ can be seen in that most textual combinations of **Oper**$_1$ (S) and S can be replaced with a single verb (see the columns 1 and 3 in Table 2) which is equal in its meaning to the noun S: *to provide aid = to aid, to ask a question = to ask*, etc. Such transformations are called *paraphrasing*. They play a significant role in theory and practice of linguistics.

2.3.2 Lexical Functions Caus and Liqu

The meaning of the LF **Caus** is 'to cause', 'to make the situation existing'. The grammatical subject for the verb $V =$ **Caus** (L) should be different from L, while the complement should be L: 1st actant of $V \neq L$; 2nd actant of $V = L$. E.g., to cause the situation of rendering attention of somebody, it is necessary to *attract* the attention of the person. Informally, **Caus** (L) answers the question: *What one does with L to cause it?*

The meaning of the LF **Liqu** is 'to eliminate', 'to cause the situation not existing'. Informally, **Liqu** $\approx$ **Caus** *not*; **Liqu** (L) answers the question: *What one does with L to eliminate it?* As for **Caus**, the grammatical subject for $V =$ **Liqu** (L) should be different from L while the complement should be L: 1st actant of $V \neq L$; 2nd actant of $V = L$. For example, to eliminate the situation of rendering attention it is necessary to *distract* this attention.

Table 3 illustrates both lexical functions.

Table 3 Examples of Caus and Liqu

LF argument	Caus	Liqu
attention	*attract*	*distract, divert*
rest	*give*	*deprive*
rights	*grant*	*deprive, concede*
pain	*give*	*calm*
floor$_2$	*give*	*deny*
parliament	*convene, convoke*	*disband, dissolve*
opportunity	*raise, afford*	*exclude, rule out*
visa	*grant, issue*	*deny, cancel*

2.3.3 Lexical Functions for Semantic Derivates

The LF **A**$_0$ is defined on nouns, verbs, and adverbs and gives the meaning equal to that of its argument, but expressed by an **adjective**: **A**$_0$ (*city*) = *urban*, **A**$_0$ (*brother*) = *fraternal*, **A**$_0$ (*move*) = *moving*, **A**$_0$ (*well*) = *good*. These are examples of semantic derivation. One could see that, in contrast to morphological word derivation, semantic derivation preserves the meaning though can take quite a different stem for the derivate.

Recall that the subscript "0" refers to the lexeme L itself rather than to any of its actants. The LFs A_1, A_2, ... , A_n, express the typical adjective qualifiers for the first, second, etc., actants of L:

x	$A_1(x)$	$A_2(x)$
surprise	*surprising*	*surprised*
buy	*buying*	*bought*

Similarly, S_0, S_1, ... , S_n, express the substantive semantic derivates, i.e., **nouns**. The function S_0 gives a noun with the same meaning as its argument, e.g., $S_0(real) = realty$, $S_0(defend) = defense$. The functions S_1, S_2, S_3, ... , S_n give the standard names of the first, second, third, nth, actants of the argument. Table 4 presents some examples for the verbs with different number of actants.

There are several substantive LFs with the meaning of standard circumstances: S_{loc} denotes a standard place, S_{mod} denotes a standard manner, etc.: $S_{loc}(to\ live) = dwelling/habitation$; $S_{mod}(to\ live) = way\ (of\ life)$, $S_{mod}(to\ write) = style$, $S_{mod}(to\ speak) = pronunciation/articulation$.

Table 4 Noun derivatives

x	$S_0(x)$	$S_1(x)$	$S_2(x)$	$S_3(x)$	$S_4(x)$
to rain	*rain*	–	–	–	–
to live	*life*	–	–	–	–
to swim	*swimming*	*swimmer*	–	–	–
to possess	*possession*	*possessor*	*property*	–	–
to love	*love*	*lover*	*beloved*	–	–
to send	*sending*	*sender*	*mail*	*recipient*	–
to buy	*buying*	*buyer*	*goods*	*seller*	*price*

In the same way, **verbal** semantic derivates are expressed by the function V_0: $V_0(phone) = to\ phone$, $V_0(frost) = to\ freeze$.

Adverbial semantic derivates are expressed by the function Adv_0: $Adv_0(rapid) = rapidly$, $Adv_0(attention) = attentively$.

There are some LFs rather similar to derivates. The function **Pred** of a noun or an adjective gives the standard **predicative verb** for its argument: **Pred** (*teacher*) = *to teach*, **Pred** (*joint*) = *to join/unite/associate*. The function **Copul** of a noun gives the standard **copulative verb** for this noun: **Copul** (*techer*) = *to be*: *John is a teacher*; **Copul** (*difficulty*) = *to present*: *This issue presents a great difficulty*; **Copul** (*help*) = *to offer*: *It will be better to offer them help and security*. These functions are used in the transformation rules that will be discussed further in Section 2.5.4, see, e.g., Rule 5. Informally, both functions answer the question "*how to be an L*," but **Copul** requires the word L as a complement, while **Pred** gives the complete answer by itself.

2.3.4 Lexical Functions for Semantic Conversives

As we have seen, there are functions related to different actants of the argument; the number of actant is indicated as a subscript of the function name. Similarly, there are functions related to two different actants of the argument; their names are given two subscripts or indices.

A **conversive** to a lexeme L is a lexeme L' which denotes the same situation from another "point of view," i.e., with a particular permutation of the actants. The application of LF **Conv**$_{ij}(L)$ means that the actant i of L becomes the actant j of L' and the actant j of L becomes the actant i of L', i.e., the actants are swapped. In general, the permutations may be possible for any number of actants, the number of possible options grows with the number of available actants, and the function name may have, accordingly, three or more indices, e.g., **Conv**$_{123}$.

The simplest examples of conversives are the words with two actants: **Conv**$_{12}$ (*parents*) = *children*, **Conv**$_{12}$ (*children*) = *parents*. The same situation can be expressed by the phrases *John and Susan are the parents of these children* or *These are the children of John and Susan*. The situation is the same, but it is presented from the "point of view" either of *John or Susan* or of the children. In some cases, the conversives are antonymous: *Peter is stronger than Dan* vs. *Dan is weaker than Peter*.

More complicated examples are verbs *to buy* and *to sell* with four actants each: **Conv**$_{31}$ (*to buy*) = *to sell*, **Conv**$_{31}$ (*to sell*) = *to buy*, see Table 5. Note that the standard names of the two corresponding actants of *to sell* are those permuted for *to buy*. Again, the same situation is presented from the "point of view" of the seller or buyer.

Table 5 Conversives

x	$S_0(x)$	$S_1(x)$	$S_2(x)$	$S_3(x)$	$S_4(x)$
to buy	*buying*	*buyer*	*goods*	*seller*	*price*
to sell	*sale*	*seller*	*goods*	*buyer*	*price*

2.3.5 Several Other Lexical Functions

There are many other lexical functions of interest, of which we will mention here the following:

- The *synonyms* are expressed through LF **Syn**. There are various degrees, or types, of synonymy, so there are several options for this LF: an absolute synonym **Syn**, a narrower synonym **Syn**$_\subset$, a broader synonym **Syn**$_\supset$, and an intersecting synonym **Syn**$_\cap$. Here are some examples: **Syn** (*digit*) = *figure*, **Syn**$_\subset$ (*respect*) = *to consider*, **Syn**$_\supset$ (*to love*) = *to adore*, **Syn**$_\cap$ (*to avoid*) = *to elude*.

- The contrastive terms, or ***antonyms***, are expressed through LF **Contr**: **Contr** (*upward*) = *downward*, **Contr** (*good*) = *bad*, **Contr** (*more*) = *less*.

- The generic terms, or ***hypernyms***, are expressed through LF **Gener**: **Gener** (*joy*) = *feeling*, **Gener** (*television*) = *mass media*.

- The standard terms for collectivity are expressed through LF **Mult**: **Mult** (*ship*) = *fleet*, **Mult** (*horse*) = *herd*, **Mult** (*goose*) = *flock*.

- The standard terms for singleness are expressed through LF **Sing**: **Sing** (*rain*) = *drop*, **Sing** (*fleet*) = *ship*, **Sing** (*gang*) = *gangster*.

- The ability to be the i-th actant in a situation is expressed through LF **Able**$_i$: **Able**$_1$ (*to read*) = *readable*, **Able**$_1$ (*to bend*) = *flexible*, **Able**$_1$ (*to eat*) = *hungry*, **Able**$_2$ (*to eat*) = *edible*, **Able**$_1$ (*to excuse*) = *apologetic*, **Able**$_2$ (*to excuse*) = *excusable*.

2.4   Lexical Functions and Collocations

As it was mentioned in the beginning, lexical function is a concept used to systematically describe "institutionalized" lexical relations, both paradigmatic and syntagmatic. Respectively, paradigmatic and syntagmatic lexical functions are distinguished. Since we are interested in collocational relations, i.e. syntagmatic institutionalized lexical relations, only syntagmatic lexical functions will be considered in this work.

2.4.1   Syntagmatic Lexical Functions

Now we provide a definition of syntagmatic lexical function from (Wanner, 2004). **A syntagmatic LF** is a (directed) standard abstract relation that holds between the base A and the collocate B of the collocation A $\oplus$ B and that denotes 'C' $\in$ 'A $\oplus$ C' with 'A $\oplus$ C' being expressed by A $\oplus$ B. 'Directed' means that the relation is not symmetric. 'Standard' means that the relation apples to a large number of collocations. 'Abstract' means that the meaning of this relation is sufficiently general and can therefore be exploited for purposes of classification.

Alongside with formalizing semantic information, lexical functions specify syntactic patterns of collocations. For this purpose, subscripts are used with the names of LFs as explained above. Subscripts identify syntactic functions of words denoting basic thematic roles associated with LF argument. We will not take semantic roles into account in our research and will treat subscripts as a part of LF name. LF names serve as tags with which collocations are to be annotated as a result of classification.

LFs can be grouped according to parts of speech to which collocational components belong. The following classes of collocations are distinguished:

1. Verb-noun: *to make a decision, to take a walk*
2. Noun-noun: *heart of the desert, prime of life*
3. Adjective-noun: *infinite patience, strong tea*
4. Adverb-verb: *to laugh heartily, to walk steadily*
5. Noun-preposition: *on the typewriter, by mail*
6. Verb-preposition: *to fly by [plane], to go to [the park]*

About 20 of standard simple LFs capture verb-noun collocations. Simple LFs can further combine to form complex LFs. Complex LFs reflect compositional semantics of collocates. For example,

$$\text{Magn} + \text{Oper}_1(laughter) = to\ roar\ [with\ laughter],$$

where *roar* means **do** [= Oper$_1$] **big** [= Magn] **laughter**. In complex LFs, or in a configuration of LFs, the syntactically central LF which determines the part of speech of the configuration and the value is written rightmost. In the above example the value is a verb, so **Oper** is written rightmost in the configuration:

$$\text{MagnOper}_1(laughter) = to\ roar\ [with\ laughter].$$

2.4.2 Syntagmatic Lexical Functions as a Collocation Typology

Collocation definition has been a controversial issue for a number of years. Various criteria have been suggested to distinguish collocations from free word combinations, they were considered in Section 2.2. Definitions based on statistical distribution of lexical items in context cover frequently encountered collocations but such collocations as feeble imagination are overlooked since they occur rarely in corpus and thus are not considered collocations in the statistical sense. On the other hand, collocation definition which suggest arbitrariness of lexical choice of the collocate depending on the base does not encompass such phrases as strong man and powerful neighbor which are considered recurrent free combinations.

We are interested in a collocation classification that can give an insight in the collocational meaning. Syntagmatic lexical functions have glosses which represent a semantic component added to the collocational base to form the meaning of a collocation. A gloss is an element of meaning found common in a group of collocations. Such groups of collocations form classes, each of the classes is associated with a particular lexical function. For verb-noun collocations, 20 lexical functions are identified. Compared to the only other existing semantic and syntactic typology, the one proposed by (Benson *et al.*, 1986), which includes 8 types of grammatical collocations and 7 types of lexical collocations, the LF typology is very fine-grained. Verb-noun lexical functions are described and exemplified in Section 5.1.

2.5 Lexical Functions in Applications

Lexical functions have the following important properties with respect to applications:

1. LFs are universal. It means that a significantly little number of LFs (about 70) represent the fundamental semantic relations between words in the vocabulary of any natural language (paradigmatic relations) and the basic semantic relations which syntactically connected word forms can obtain in the text (syntagmatic relations).

2. LFs are idiomatic. LFs are characteristic for idioms in many natural languages. An example is the lexical function **Magn** which means 'a great degree of what is denoted by the key word'. In English, it is said *to sleep soundly* and *to know firmly,* not the other way round: **to sleep firmly* or **to know soundly*. But in Russian the combination *krepko spat'* (literally *to sleep firmly*) is quite acceptable although it is not natural in English.

3. LFs can be paraphrased. For example, the LFs **Oper** and **Func** can form combinations with their arguments which are synonymous to the basic verb like in the following utterances: *The government controls prices – The government has control of prices –* The *government keeps prices under control – The prices are under the government's control.* Most paradigmatic lexical functions (synonyms, antonyms, converse terms, various types of syntactic derivatives) can also substitute for the keyword to form synonymous sentences.

4. LFs are diverse semantically. Sometimes the values of the same LF from the same key word are not synonymous. This is especially characteristic of the LF Magn. We can describe a great degree of *knowledge* in the following three ways: a) as *deep* or *profound*; b) as *firm*; c) as *broad* or *extensive*. Although all these adjectives are valid values of the **Magn**, the three groups should somehow be distinguished from each other because the respective adjectives have very different scopes in the semantic representation of the keyword. *Deep* and *profound* characterize *knowledge* with regard to the depth of understanding; *firm* specifies the degree of its assimilation; *broad* and *extensive* refer to the amount of acquired knowledge. It was proposed in (Apresjan *et al.*, 2003) that in order to keep such distinctions between different values of the same LFs in the computerized algebra of LFs it is sufficient to ascribe to the standard name of an LF the symbol NS (non-standardness) plus a numerical index and maintain the correspondences between the two working languages by ascribing the same names to the respective LFs in the other language.

2.5.1 Word Sense Disambiguation

Syntagmatic LFs can be used to resolve syntactic and lexical ambiguity. Both types of ambiguity can be resolved with the help of LFs **Oper, Func** and **Real**.

LFs like **Magn** (with the meaning 'very', intensifier), **Bon** (with the meaning 'good'), **Figur** (with the meaning 'metaphoric expression typical of the key word') among others, can be used to resolve lexical ambiguity.

2.5.1.1 Syntactic Ambiguity

Syntactic ambiguity and its resolution with the help of LFs is explained in this section by means of an example. Let us consider such phrases as *support of the parliament* or *support of the president.* The word *support* is object in the first phrase, but it is subject (agent) in the second phrase. Syntactically, both phrases are identical: *support* + the preposition *of* + a noun, this is the sourse of syntactic ambiguity, and for that reason both phrases may mean both: 'support given by the parliament (by the president)', which syntactically is the subject interpretation with the agentive syntactic relation between *support* and the subordinated noun, and 'support given to the parliament (to the president)' which syntactically is the object interpretation with the first completive syntactic relation between *support* and the subordinated noun.

This type of ambiguity is often extremely difficult to resolve, even within a broad context. LF support verbs can be successfully used to disambiguate such phrases because they impose strong limitations on the syntactic behaviour of their keywords in texts.

Now let us view the same phrases in a broader context. The first example is *The president spoke in support of the parliament,* where the verb *to speak in* is **Oper₁** the noun *support.* Verbs of the **Oper₁** type may form collocations with their keyword only on condition that the keyword does not subordinate directly its first actant. The limitation is quite natural: **Oper₁** is by definition a verb whose grammatical subject represents the first actant of the keyword. Since the first actant is already represented in the sentence in the form of the grammatical subject of **Oper₁**, there is no need to express it once again. This is as much as to say that the phrase *The president spoke in support of the parliament* can only be interpreted as describing the support given to the parliament, with *parliament* fulfilling the syntactic function of the complement of the noun *support.*

On the other hand, verbs of the **Oper₂** type may form such collocations only on condition that the keyword does not subordinate directly its second actant. Again, the limitation is quite natural: **Oper₂** is by definition a verb whose grammatical subject represents the second actant of the keyword. Since the second actant is already represented in the sentence in the form of the grammatical subject of **Oper₂**, there is no need to express it once again. So in the second example we consider in this section, *The president enjoyed* (**Oper₂**) *the support of the parliament,* the phrase *the support of the parliament* implies the support given to the president by the parliament, with *parliament* fulfilling the syntactic function of the agentive dependent of the noun *support.*

In cases of syntactic ambiguity, syntactically identical phrases are characterized by different lexical functions which in this case serve as a tool of disambiguation.

2.5.1.2 Lexical Ambiguity

LFs are also useful in resolving lexical ambiguity. For the sake of brevity, we shall give only one illustrative example. The Russian expression *provodit' razlichie* and its direct English equivalent *to draw a distinction* can be analyzed as composed of **Oper$_1$** + its keyword. Taken in isolation, the Russian and the English verbs are extremely polysemous, and choosing the right sense for the given sentence becomes a formidable problem. *Provodit'*, for example, has half a dozen senses ranging from 'spend' via 'perform' to 'see off', while *draw* is a polysemic verb for which dictionaries list 50 senses or more. However, in both expressions the mutual lexical attraction between the argument of the LF and its value is so strong that, once the fact of their co-occurrence is established by the parser, we can safely ignore all other meanings and keep for further processing only the one relevant here.

2.5.2 Computer-Assisted Language Learning

One of the purposes in developing software for language learning is to acquire lexicon. It has been proposed (for example, in (Diachenko, 2006)) to organize learning in the form of linguistic games. There are games that operate on a word dictionary, but in order to learn collocations, a lexical function dictionary can be used whose advantage is that it includes the linguistic material on word combinations which is absent in word dictionaries. Below an example of a game oriented to lexical functions is given.

2.5.2.1 Game "Lexical Function"

In the game "LF", the user needs to supply values of a concrete LF for each given argument. The user chooses the LF she is going to play with and enters the number of questions. The system gathers the material for the game by random selection of arguments from the dictionary. The system also shows the user the definition of the LF and two examples.

According to the difficulty of learning, all LFs were divided into 3 levels. While some LF values have a compound format and may include the argument itself or pronouns, the system generates hints for such values. For example, **CausFact$_0$** (with the meaning 'to cause something to function according to its destination) for *clock* (in the sense of 'time-measuring instrument) is *wind up (a watch) / start (a watch)*. For this value the hint will look like this:

"— (*a watch*)"

CausFact₀ of *imagination* is *fire* (somebody's *imagination*). The hint will look like this

"— (somebody) (*imagination*)".

If the user cannot supply an answer, the system shows him or her the list of correct answers.

2.5.3 Machine Translation

Two important properties of LFs mentioned in the beginning of Section 2.5, i.e., their semantic universality and cross-linguistic idiomaticity, make them an ideal tool for selecting idiomatic translations of set expressions in a MT system. The way it can be done is explained by an example of prepositions in English and Spanish.

As is well known, locative prepositions used to form prepositional phrases denoting places, sites, directions, time points, periods, intervals etc. reveal great versatility within one language and incredibly fanciful matching across languages. If we were to account properly for the discrepancies existing between the uses of these prepositions, say, in English and Spanish, we would have to write too detailed translation rules involving complicated semantic and pragmatic data. However, a large part of the task may be achieved with the help of LFs.

Consider the following correspondences between English and Spanish that may be easily found with the help of the LFs **Dir** (preposition denoting a movement toward the location expressed by the keyword):

Dir(*city*) = *to* (*the city*), **Dir**(*cuidad*) = *a* (*la cuidad*),
Dir(*friend*) = *to* (*my friend*), **Dir**(*amiga*) = *con* (*mi amiga*).

In order to ensure the production of these equivalents in machine translation, we must only identify the arguments and the value of the LF during parsing and substitute the correct value from the target language dictionary during generation.

2.5.3.1 Implementation Example

A module that annotates word combinations with lexical functions (such word combinations will be collocations) represented in Fig. 1, can be included in any machine translation system based on interlingua like UNL. UNL is a project of multilingual personal networking communication initiated by the University of United Nations based in Tokyo (Uchida *et al.*, 2006).

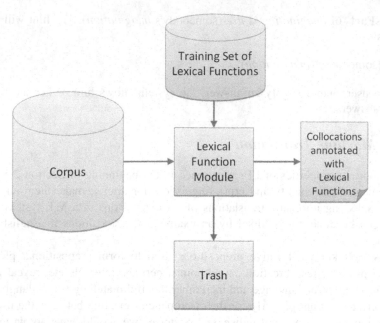

Fig. 1 Lexical Function Module of a Machine Translation System.

2.5.4 Text Paraphrasis

Lexical functions present a powerful mechanism for synonymous transformations (paraphrasing) applicable on deep syntactic or semantic levels. The corresponding bi-directional formulae (equivalencies, $\Leftrightarrow$) contain small subtrees on both sides; in some cases a subtree on one side or both is reduced to a single node.

The transformations work on subtrees of the whole syntactic tree or semantic network for a sentence. If the whole tree contains a subtree matching the one on either side of the formula, then it can be substituted with the tree on the other side of the rule. For any rule and any direction of its use, the meaning of the whole tree remains unchanged.

Here are several examples. They are related to some word that is traditionally denoted by C_0, and its syntactic surrounding. The marks on the arrows correspond to the different types of syntactic relations (1, ..., 5 stand for actants, in particular, 2 stands for the direct object; *attr* is for attributive relation).

1. The synonyms are equal in their meaning:

   $C_0 \Leftrightarrow \mathbf{Syn}\,(C_0)$.

   Example: *easy* $\Leftrightarrow$ *simple*.

2. A combination of $\mathbf{Oper_1}\,(N)$ with a dependent noun N, where N is equal in its meaning to some verb C_0, i.e., $N = \mathbf{S_0}\,(C_0)$, is reducible to this verb C_0:

   $\mathbf{Oper_1}\,(\mathbf{S_0}\,(C_0)) \xrightarrow{\;2\;} \mathbf{S_0}\,(C_0) \Leftrightarrow C_0$.

Example: *to offer help* ⟺ *to help.*

3. A combination of **Gener** (C_0) with a dependent noun C_0 is reducible to C_0:

   Gener (C_0) $\xrightarrow{2}$ C_0 ⟺ C_0.

   Example: *a feeling of happiness* ⟺ *happiness.*

4. A combination of **Gener** attributed by A_0 is reducible to their common argument:

   Gener (C_0) $\xrightarrow{attr}$ A_0 (C_0) ⟺ C_0.

   Example: *an explosive material* ⟺ *an explosive.*

5. A copulative word combination is reducible to LF **Pred:**

   Copul (C_0) $\xrightarrow{2}$ C_0 ⟺ **Pred** (C_0).

   Example: *to be a teacher* ⟺ *to teach.*

6. A verb can be replaced by its conversive:

$$L \begin{array}{c} \nearrow^{i} P \\ \searrow_{j} Q \end{array} \quad \Leftrightarrow \quad \mathbf{Conv}_{ij}\,(L) \begin{array}{c} \nearrow^{i} Q \\ \searrow_{j} P \end{array} \quad , \; i,j = 1, \ldots, 5; \; i \neq j.$$

Example: *John bought the book from Peter at ten dollars* ⟺ *Peter sold* [= **Conv**$_{13}$ *(to buy)*] *the book to John at ten dollars.*

The total number of such rules known in the Meaning-Text theory reaches several tens. They can be implemented through about thirty standard operations, mainly connected with tree transformations: renaming of a node in a tree, elimination or insertion of a node, moving an arc to another node, etc. A detailed discussion of the mechanism of application of these rules can be found in (Wanner, 1996; Mel'čuk, 1974; Steel, 1990).

2.5.4.1 Applications of the Lexical Functions and Paraphrastic Rules

Lexical functions in various ways can be used in linguistic transformations in text analysis, synthesis, and translation.

- Through the use of paraphrases for single node or nodes constituting a syntactic sybtree, one can significantly decrease the variety of lexemes occurred in a discourse. Thus, the semantic representation becomes more homogeneous and the problem of "understanding" the discourse is facilitated. See Rule 1 or 6 as examples.

- Through the use of paraphrases diminishing the number of nodes in the deep syntactic structure, one can decrease the total number of nodes in the structure. Since the rules of logical inference are usually concise, it is easier to match more compact structures with them. See Rules 2 to 5 as examples.
- In text synthesis, through the use of lexical functions, the correct words for standard meanings can be chosen for a given language. For example, to represent the meaning of 'high intensity' with the concept 'tea', the words *strong*, *starker* ('powerful'), *cargado* ('loaded'), *fort* ('forceful'), *krepkiy* ('firm'), or *mocna* ('firm') can be chosen for English, German, Spanish, French, Russian, and Polish, correspondingly. We are not aware of any elegant method of choosing the word *cargado* for the meaning of 'high intensity' in the context of *té*, other than the use of the lexical function **Magn**.
- Through the use of lexical functions on the deep syntactic level in translation from one natural language to another, one can easily obtain quite idiomatic translations. This is illustrated by the following two examples of English-to-Spanish translations:

1. *strong tea* $\Rightarrow$ **Magn** (*tea*) $\xleftarrow{\quad attr \quad}$ *tea* $\Rightarrow$ *té* $\xrightarrow{\quad attr \quad}$ **Magn** (*té*) $\Rightarrow$ *té cargado*.

2. *asks questions* $\Rightarrow$ **Oper**$_1$ (*question*)$_{\text{sing, 3pers}}$ $\xrightarrow{\quad 2 \quad}$ *question*$_{\text{pl}}$ $\Rightarrow$ **Oper**$_1$ (*pregunta*)$_{\text{sing, 3pers}}$ $\xrightarrow{\quad 2 \quad}$ *pregunta*$_{\text{pl}}$ $\Rightarrow$ *hace preguntas*.

Notice that LFs **Magn** and **Oper**$_1$ remain invariant during interlingual transfer in the previous examples, like elements of some universal language. The labeled syntactic structures involved are also applicable to both languages. As to the lexical expressions corresponding to these LFs in various languages, they can be very specific. Thus, though it is not possible to directly translate the function values, it is possible to share the functions across languages. Even if some LFs are not meaningful by themselves, they can serve as elements of an intermediate language of a very deep, though not semantic, level.

Chapter 3
Identification of Lexical Functions

3.1 Lexical Functions in Dictionaries

In this section we describe a number of lexical resources that contain lexical functions. Actually, they are not specialized dictionaries of lexical functions only, but include LFs together with other linguistic information. However, in the resources presented here, the LF apparatus plays a strategic role in how the lexical material is selected, organized and structured, and this fact distinguished these word repositories from conventional dictionaries.

As it was mentioned in Chapter 2, the concept of lexical function was originally proposed by researchers of the Russian semantic school within the context of the Meaning-Text Theory. Lexical functions are applied there for describing combinatorial properties of lexica, i.e., how words are grouped in natural language texts to convey certain meanings. At the same time, the LF formalism serves as a practical instrument in language applications.

It is widely acknowledged that a computer implementation of a linguistic theory, if successful, can play the role of an independent "judge" whose opinion is objective and free of subjectivity. If a computer program developed on the basis of a particular linguistic theory fulfills its task effectively, then the linguistic theory is verified to be true. To prove the Meaning-Text Theory, the sample implementation was chosen to be a machine translation system.

As a result, three versions of such system have been developed within almost 30 years. The system is called ETAP, the name is built with the first letters of the Russian phrase which can be translated in English as "Electro-Technical Automatic Translation" (*Translation* is *Perevod* in Russian). The last version, ETAP-3, was developed in the Laboratory of Computational Linguistics of Kharkevich Institute of Information Transmission Problems, Russian Academy of Sciences. This machine translation (MT) system is based on rules designed on the principles of the Meaning-Text Theory. It works with two natural languages, English and Russian, and with an artificial language, UNL, referred to in Section 2.5.3 (Uchida *et al.*, 2006). ETAP-3 MT System can be accessed online at http://cl.iitp.ru/etap.

Dictionary is a vital part of a rule-based machine translation system. For ETAP, the dictionary called the *Explanatory Combinatorial Dictionary* was compiled

A. Gelbukh, O. Kolesnikova: Semantic Analysis of Verbal Collocations, SCI 414, pp. 41–50.
springerlink.com   © Springer-Verlag Berlin Heidelberg 2013

where lexical functions were used a mechanism for a systematic and consistent treatment of collocational properties of individual words. Now we are going to give a more detailed description of this dictionary.

3.1.1 Explanatory Combinatorial Dictionary

This dictionary was first compiled for Russian (Mel'čuk and Zholkovskij 1984) and for French (Mel'čuk *et al.* 1984, 1988). It is **explanatory** in the sense that for every lexical unit, it gives its form, meaning and usage. Examples of other explanatory dictionaries are the *Concise Oxford Dictionary of Current English*; Merriam Webster dictionaries; *Collins COBUILD English Language Dictionary* (COBUILD – Collins Birmingham University International Language Database).

Collocational or combinatorial properties become apparent in typical collocations where a given word functions as the headword. It means that they describe a word's environment, and thus characterize its usage. Classical **usage dictionaries**, for example, *Usage and Abusage: A Guide to Good English* (Partridge, 1994), explain word differences, indicate what word usage sounds correct and natural, specify restrictions on the usage of certain words, and list certain typical constructions.

In our opinion, the Explanatory Combinatorial Dictionary (ECD) is best described in (Mel'čuk, 2006). ECD is as a formalized semantically-based lexicon designed to be a part of a linguistic model, in particular, the model build on the principles of the Meaning-Text Theory.

Each entry includes full information of a word or idiom treated in a non-compositional manner as a single lexical unit. Word characteristics and properties are described in a consistent way covering various aspects: semantic (word definition and connotations), phonetic, graphical and combinatorial. The combinatorial data includes morphological information (inflexions with which a given word combines), syntactic valency, lexical functions, the word's usage in different language styles and dialects, its usage in real life situations (pragmatic information) and examples to make the entry illustrative.

Since LFs are of our interest, here we offer a partial example of the ECD entry for the verb *to bake* in the meaning "to cook food", in the part of data on lexical functions. In this example, we list some LFs typical for *to bake*. $S_{instr-loc}$ denotes a typical instrument and a place of *to bake*, **Ver** (from Lat. *verus*, real, genuine) means "as it should be", "meeting intended requirements"; **AntiVer** is the opposite of **Ver**.

> *BAKE*
> **Syn**:   *make baked* (something)
> **Syn$_\subset$**:   *cook*1
> **Syn$_\cap$**:   *roast*1
> **S$_{instr-loc}$**:   *oven*
> **Ver**:   *to a turn*
> too much, **AntiVer**: *overbake*
> not enough, **AntiVer**: *underbake*

Even this little example shows how semantic derivatives and collocations can be ordered and represented in a consistent way using the LFs apparatus. Once we list all LFs for a particular word, we get access to complete semantic and syntactic information concerning this word.

3.1.2 Combinatorial Dictionaries for Machine Translation

The dictionaries used in ETAP-3 machine translation system are called **combinatorial** to emphasize the fact that their purpose is to be repositories of information on typical collocations and co-occurrence preferences encoded as lexical functions. Combinatorial dictionaries in ETAP are the abridged versions of explanatory combinatorial dictionaries of the Meaning-Text Theory. In Russian-to-English translation, the English combinatorial dictionary is used as the target dictionary, and as the source dictionary in English-to-Russian translation. Both dictionaries (English and Russian) contain about 65,000 lexical entries each and offer rich information on syntactic and semantic features of the word, its government pattern, and values of lexical functions for which the word is the argument. Lexical entries sometimes include parsing and transfer rules and pointers to these rules (Apresian *et al.*, 2003).

Each entry in combinatorial dictionaries is organized as a set of sections or zones. The first section is universal and includes data independent of any language application. All other sections provide machine translation specific information. Here we will present examples of two lexical entries from (Iomdin and Cinman, 1997). The way the entries for *applause* and *respect* are shown is different from the actual entries in the combinatorial dictionaries because these lexicons were designed for machine use only. Here we represent only the section of co-occurrence data, i.e., the LFs values for the arguments *applause* and *respect*, and organize them in tables. The information is also displayed in a human readable form, adding the LF meanings for the reader's convenience.

APPLAUSE		
LF	**Meaning**	**LF values**
Syn	synonym	*ovation*
V₀	verbal semantic derivative	*applaud*
Magn	intense, very, intensifier	*heavy/lengthy/prolonged/loud1/ thunderous/ringing/terrific*
AntiMagn	little, the opposite of intense	*light2/weak/thin*
Ver	true, genuine, correct, in accord with its purpose, such as it should be	*well-merited*
Mult	multiplicity, plurality, typical name of a collection, a multitude of something	*storm/round1/thunder1*

Oper$_1$	carry out	*give*
Oper$_2$	undergo	*draw1/win/get*
Labor$_{12}$	support verb which connects the agent as the subject with the patent as the direct object, the argument functions as indirect object	*meet<with>/greet<with>/ hail2<with>/rise1<in1>*
IncepFunc$_0$	the argument begins to take place	*break1<out>*
FinFunc$_0$	the argument is terminated	*subside/die<out>*

RESPECT		
LF	**Meaning**	**LF values**
Oper$_1$	carry out	*have*
Oper$_2$	undergo	*command2*
IncepOper$_2$	begin to undergo	*win*
Labor$_{12}$	support verb which connects the agent as the subject with the patent as the direct object, the argument functions as indirect object	*hold1<in1>*
Manif	manifest, demonstrate, make explicit	*show*
Ver	true, genuine, correct, in accord with its purpose, such as it should be	*genuine*
Magn	intense, very, intensifier	*high1/profound/immense/solid1*
AntiMagn	little, the opposite of intense	*scant1*

It can be easily seen that the LFs data is represented in the same way as in the Explanatory Combinatorial Dictionary described in the previous subsection. The combinatorial dictionaries in ETAP-3 have acquired great importance because of their uniqueness. In particular, the English combinatorial dictionary for ETAP-3 is, to our knowledge, the only dictionary of such kind for English.

The reader may compare the entries for *applause* and *respect* in the combinatorial dictionaries of ETAP-3 with the entries for the same nouns in the *BBI Combinatory Dictionary of English* (Benson *et al.*, 1986) given below.

APPLAUSE

n. 1. to draw, get, win ~ for
2. heavy, lengthy, prolonged; light, weak; loud, thunderous ~
3. a burst; ripple; round of ~
4. to (the) ~ (she appeared on stage to the thunderous ~ of her admirers)

RESPECT

I n. ['esteem'] 1. to pay, show ~ to
2. to command, inspire ~ (she commands ~ from everyone = she commands everyone's ~)
3. to earn, win; lose smb.'s ~
4. deep, profound, sincere; due; grudging; mutual ~
5. ~ for (~ for the law)
6. out of ~ (she did it out of ~ for her parents)
7. in ~ (to hold smb. in ~)
8. with ~ (with all due ~, I disagree) ['regard']
9. in a ~ (in this ~; in all ~s)
10. in, with ~ to

It can be observed immediately that the collocational information in the BBI Combinatory Dictionary of English is more complete than the data in the combinatorial dictionary of ETAP-3. But the advantage of the latter is that all collocations are semantically annotated, therefore the reader, who is very often a learner of English as a second language, can get a clear idea of what a given collocation means. The first dictionary provides a list of typical constructions ordered according to their syntactic patterns without giving details on semantics, so the reader has to consult some other dictionary to get a full understanding of what the collocations she has just learnt mean.

3.1.3 French Combinatorial Dictionaries

The descriptive principles of the Explanatory Combinatorial Dictionary were applied in compiling the DiCo, a computerized formal database of French semantic derivatives and collocations, and the *Lexique actif du français* (LAF, 'Active lexicon of French'). The latter was entirely generated from the DiCo, therefore, presents the same material, but with the objective to make it comprehensible for general public (Polguère, 2000; 2007).

As to the linguistic contents, **DiCo** includes entries of those French words that present a special challenge in natural language processing due to their complex combinatorial properties. Concerning the DiCo format, it was designed as platform-independent as possible. At the same time, this lexical database has a certain relation to the English WordNet (Miller, 1998), because some lexical functions in fact represent particular lexical relations captured in WordNet, for example, the relation of synonymy is encoded as the lexical function **Syn**, and hypernyms are represented as values of the LF **Gener** (from Lat. *genus*, the closest generic concept for the argument). More information can be found at http://olst.ling.umontreal.ca/dicouebe where DiCo can be accessed and searched.

The LAF dictionary is a popularized version of DiCo. It contains data of about 20,000 French semantic derivations and collocations and was developed in such a way as to make the description of lexica accessible to non-linguists, mainly, to language teachers and students.

3.1.4 Dictionary of Spanish Collocations

Diccionario de colocaciones del español (DiCE), a web-based dictionary of Spanish collocations (Alonso *et al.*, 2010; Vincze *et al.*, 2011), contains words annotated with lexical functions used as a typology to structure semantic and combinatorial information on lexical units. The dictionary can be accessed online at http://www.dicesp.com/paginas.

DiCe is compiled for language researchers but is not limited to scientific purposes. Its design also allows general public, including learners of Spanish as a second language, to take advantage of valuable information on collocations. LF annotations are presented in the form they were elaborated within the Meaning-Text Theory (Mel'čuk, 1996), which is relevant for experts in linguistics. At the same time, all the information encoded with the help of the LF formalism is explained in a simple way, by means of definitions called natural language glosses, by analogy with WordNet glosses (Miller, 1998).

Currently, the word list of DiCe includes only lexical units belonging to the semantic field of emotions, and the dictionary specifies about 19,500 lexical relations as LF values.

Each word presented as a lexical unit in DiCe is described with respect to its semantics and combinatorial properties. **Semantic** information includes generic meaning of the word, semantic roles of words commonly used with the given word in speech, corpus examples as illustrations, quasi-synonyms and quasi-antonyms of the word. **Combinatorial** information includes details on the word's subcategorization frame and its typical collocations and/or semantic derivatives annotated with lexical functions. Vincze *et al.* (2011) present a sample entry from the DiCe and explain its structure and annotations.

3.2 Automatic Identification of Lexical Functions

Research on automatic detection of lexical functions is a recently emerged area of natural language processing. Wanner (2004) and Wanner *et al.* (2006) report results on performance of a few machine learning algorithms on classifying collocations according to the typology of lexical functions. In this section, we will summarize the work done in (Wanner, 2004) and (Wanner *et al.*, 2006), then we will also mention another research on automatic extraction of lexical functions (Alonso Ramos *et al.*, 2008) based on an approach different from the work in

(Wanner, 2004) and (Wanner *et al.*, 2006). The section is concluded with three statements, or hypotheses, made in (Wanner *et al.*, 2006).

3.2.1 Research in (Wanner, 2004) and (Wanner et al., 2006)

In 2004, L. Wanner proposed to view the task of LF detection as automatic classification of collocations according to LF typology. To fulfill this task, the nearest neighbor machine learning technique was used. Datasets included Spanish verb-noun pairs annotated with nine LFs: $CausFunc_0$, $Caus_2Func_1$, $IncepFunc_1$, $FinFunc_0$, $Oper_1$, $ContOper_1$, $Oper_2$, $Real_1$, $Real_2$.

Verb-noun pairs were divided in two groups. In the first group, nouns belonged to the semantic field of emotions; in the second groups nouns were field-independent. As a source of information for building the training and test sets, hypernymy hierarchy of the Spanish part of EuroWordNet was used. The words in the training set were represented by their hypernyms, Basic Concepts and Top Concepts. The average *F-measure* of about 70% was achieved in these experiments. The best result for field-independent nouns was F-measure of 76.58 for $CausFunc_0$ with the meaning 'cause the existence of the situation, state, etc.' The Causer is the subject of utterances with $CausFunc_0$.

In (Wanner *et al.*, 2006), four machine learning methods were applied to classify Spanish verb-noun collocations according to LFs, namely Nearest Neighbor technique, Naïve Bayesian network, Tree-Augmented Network Classification technique and a decision tree classification technique based on the ID3-algorithm.

As in (Wanner, 2004), experiments were carried out for two groups of verb-noun collocations: nouns of the first group belonged to the semantic field of emotions; nouns of the second group were field-independent. Lexical functions were also identical with (Wanner, 2004) as well as data representation.

The best results for field-independent nouns were shown by ID3 algorithm (F-measure of 0.76) for $Caus_2Func_1$ with the meaning 'cause something to be experienced / carried out / performed', and by the Nearest Neighbor technique (F-measure of 0.74) for $Oper_1$ with the meaning 'perform / experience / carry out something'. The Causer is the subject of utterances with $Caus_2Func_1$, and the Agent is the direct object of the verb which is the value of $Caus_2Func_1$. In utterances with $Oper_1$, the Agent is the subject.

As we are interested in experimented with verb-noun collocations where the nouns are have various semantics, i.e., the nouns are field-independent, Tables 8 and 9 summarize the results for field-independent nouns only in (Wanner, 2004) and (Wanner *et al.*, 2006).

Table 6 gives the meaning of lexical functions used in experiments only with field-independent nouns (Wanner, 2004), examples in Spanish with literal translation in English; #Tr stands for the number of examples of a given LF in the training set; #T stands for the number of examples of a given LF in the test set; #Tt stands for the total number of examples of a given LF in the training set and in the test set.

Table 7 lists LFs with total number of examples in (Wanner *et al.*, 2006) for verb-noun combinations with field-independent nouns. Table 8 presents the results reported in (Wanner, 2004). Table 9 shows the results in (Wanner *et al.*, 2006); the values of precision, recall and F-measure are given in the following format: <precision> | <recall> | <F-measure>. Not all four machine learning methods in Table 9 were applied to all LFs; if experiments were not made for a particular method and LF, N/A is put instead of precision, recall, and F-measure.

Table 6 Examples of data in (Wanner, 2004)

Name	Meaning	Examples in Spanish	Lit. translation in English	#Tr	#T	#Tt
$Oper_1$	experience, perform, carry out something	*dar golpe* *presentar una demanda* *hacer campaña* *dictar la sentencia*	*give a blow* *present a demand* *make a campaign* *dictate a sentence*	35	15	50
$Oper_2$	undergo, be source of	*someterse a un análisis* *afrontar un desafío* *hacer examen* *tener la culpa*	*submit oneself to analysis* *face a challenge* *make exam* *have guilt*	33	15	48
$CausFunc_0$	cause the existence of the situation, state, etc.	*dar la alarma* *celebrar elecciones* *provocar una crisis* *publicar una revista*	*give the alarm* *celebrate elections* *provoke a crisis* *publish a magazine*	38	15	53
$Real_1$	act accordingly to the situation, use as forseen	*ejercer la autoridad* *utilizar el teléfono* *hablar la lengua* *cumplir la promesa*	*exercise authority* *use a telephone* *speak a language* *keep a promise*	37	15	52
$Real_2$	react accordingly to the situation	*responder a objeción* *satisfacer un requisito* *atender la solicitud* *rendirse a persuasión*	*respond to an objection* *satisfy a requirement* *attend an application* *surrender to persuasion*	38	15	53

Table 7 Data in (Wanner *et al.*, 2006)

LF	Number of Examples
$CausFunc_0$	53
$Oper_1$	87
$Oper_2$	48
$Real_1$	52
$Real_2$	53

Table 8 Results in (Wanner, 2004)

F-measure/LF	$CausFunc_0$	$Oper_1$	$Oper_2$	$Real_1$	$Real_2$
field-independent nouns	76.58	60.93	75.85	74.06	58.32

Table 9 Results in (Wanner *et al.*, 2006)

LF	Machine learning technique			
	NN	NB	TAN	ID3
CausFunc$_0$	0.59 | 0.79 | 0.68	0.44 | 0.89 | 0.59	0.45 | 0.57 | 0.50	N/A
Caus$_2$Func$_1$	N/A	N/A	N/A	0.53 | 0.65 | 0.50
FinFunc$_0$	N/A	N/A	N/A	0.53 | 0.40 | 0.40
IncepFunc$_1$	N/A	N/A	N/A	0.40 | 0.48 | 0.40
Oper$_1$	0.65 | 0.55 | 0.60	0.87 | 0.64 | 0.74	0.75 | 0.49 | 0.59	0.52 | 0.51 | 0.50
Oper$_2$	0.62 | 0.71 | 0.66	0.55 | 0.21 | 0.30	0.55 | 0.56 | 0.55	N/A
ContOper$_1$	N/A	N/A	N/A	0.84 | 0.57 | 0.68
Real$_1$	0.58 | 0.44 | 0.50	0.58 | 0.37 | 0.45	0.78 | 0.36 | 0.49	N/A
Real$_2$	0.56 | 0.55 | 0.55	0.73 | 0.35 | 0.47	0.34 | 0.67 | 0.45	N/A

3.2.2 Research in (Alonso Ramos et al., 2008)

Alonso Ramos *et al.* (2008) propose an algorithm for extracting collocations following the pattern "support verb + object" from FrameNet corpus of examples (Ruppenhofer *et al.*, 2006) and checking if they are of the type **Oper$_n$**. This work takes advantage of syntactic, semantic and collocation annotations in the FrameNet corpus, since some annotations can serve as indicators of a particular LF. The authors tested the proposed algorithm on a set of 208 instances. The algorithm showed accuracy of 76%. The researchers conclude that extraction and semantic classification of collocations is feasible with semantically annotated corpora. This statement sounds logical because the formalism of lexical function captures the correspondence between the semantic valency of the keyword and the syntactic structure of utterances where the keyword is used in a collocation together with the value of the respective LF.

3.2.3 Three Hypothesis Stated by Wanner et al. (2006)

Wanner *et al.* (2006) experiment with the same type of lexical data as in (Wanner, 2004), i.e., verb-noun pairs. The task is to answer the question: what kind of collocational features are fundamental for human distinguishing among collocational types. The authors view collocational types as LFs, i.e. a particular LF represents a certain type of collocations. Three hypotheses are put forward as possible solutions, and to model every solution, an appropriate machine learning technique is selected. Below we list the three hypotheses and the selected machine learning techniques.

1. Collocations can be recognized by their similarity to the prototypical sample of each collocational type; this strategy is modeled by the Nearest Neighbor technique.
2. Collocations can be recognized by similarity of semantic features of their elements (i.e., base and collocate) to semantic features of elements of the collocations known to belong to a specific LF; this method is modeled by

Naïve Bayesian network and a decision tree classification technique based on the ID3-algorithm.

3. Collocations can be recognized by correlation between semantic features of collocational elements; this approach is modeled by Tree-Augmented Network Classification technique.

It should be mentioned, that having proposed three hypotheses, the authors have not yet demonstrated their validity by comparing the performance of many machine learning techniques known today, but applied only four learning algorithms to illustrate that three human strategies mentioned above are practical. This will be considered in more detail in Chapter 5.

3.3   Automatic Detection of Semantic Relations

There has been some research done on semantic relations in word combinations, for example, one that deals with automatic assignment of semantic relations to English noun-modifier pairs in (Nastase and Szpakowicz, 2003; Nastase *et al.*, 2006). Though in our work, verb-noun combinations are treated, we believe that the principles of choosing data representation and machine learning techniques for detection of semantic relations between a noun and a modifier can also be are used to detect semantic relations in verb-noun pairs. The underlying idea is the same: learning the meaning of word combinations. In (Nastase and Szpakowicz 2003, Nastase *et al.* 2006), the researchers examined the following relations: causal, temporal, spatial, conjunctive, participant, and quality. They used two different data representations: the first is based on WordNet relations, the second, on contextual information extracted from corpora. They applied memory-based learning, decision tree induction and Support Vector Machine. The highest F-measure of 0.847 was achieved by C5.0 decision tree to detect temporal relation based on WordNet representation.

Chapter 4
Meaning Representation

4.1 Linguistic Meaning and its Representation

What is meaning? What is the meaning of a word, a phrase, a text? Usually, in our everyday life, when the meaning of some word is not clear to us, we look it up in a dictionary. In this case it may be stated, that the meaning of the word we are interested in is its **dictionary definition**. Let us consider, as an example, the meaning of *extremophile* in the Merriam-Webster Online English Dictionary at http://www.merriam-webster.com:

> *extremophile*, noun : an organism that lives under extreme environmental conditions (as in a hot spring or ice cap)

It may be noticed then, that the meaning of *extremophile* or what the word *extremofile* signifies is, in fact, this very concrete object, that exact living being which is described in the previously given definition. Following this line of thinking, the dictionary definition is thus seen not as the meaning but as an **explanation** of that meaning in words other than *extremophile*, or, in more general terms, a **meaning representation**.

Another meaning representation of the same word is its encyclopedic definition in Encyclopedia Britannica at http://www.britannica.com/:

> *extremophile*, an organism that is tolerant to environmental extremes and that has evolved to grow optimally under one or more of these extreme conditions, hence the suffix "phile", meaning "one who loves".

Still another manner to look for the meaning of *extremophile* is to search the web for corresponding images. Then, we may view the photo given below as a meaning representation of the word under study (the image is retrieved from www.wikipedia.org; author: Rpgch):

A. Gelbukh, O. Kolesnikova: Semantic Analysis of Verbal Collocations, SCI 414, pp. 51–59.
springerlink.com

How can the meaning be represented for a computer? How can the meaning of one word be distinguished from the meaning of another word by a natural language application? How can a computer program compare meanings and determine if they are similar or different? And to what degree are they similar or different?

One of the approached in natural language processing is to represent word meaning with its linguistic characteristics and properties or **features**. Since language is a multi-level object and a word as a linguistic phenomenon reveals itself and functions at various levels, features are classified according to relevant levels:

Phonetics and orthography: how the word is pronounced or written, what sounds are included in the "audio image" of the word, on what sound the accent is placed, how the sounds that "make up" the word interact and adapt themselves to each other so that we can pronounce the word easily and comfortably.

Morphology: what forms the word has, what prefixes or suffixes it may take, whether the word forms are regular (formed according to well-defined rules) or irregular.

Lexis: what lexical relations the word has with other words in the language, i.e., what synonyms, antonyms, hypernyms, hyponyms, meronyms, etc. it has.

Syntax: how the word functions in phrases, what positions it can or can not take in utterances.

Semantics: what semantic role the word has, whether it denotes an agent, a patient, location, time, instrument, etc.

Pragmatics: how the word is related to intentions, emotions, beliefs, motives of the speaker.

Etymology: what origin the word has, what words of the same or other (mostly ancient) languages served as its "building bricks".

Stylistics: in texts of what styles or (poetic, fiction, scientific, technical, journalistic, etc.) the word is typically used.

Dialectology: in what geographic regions and locations the word is used, what its "natural habitat" is.

Terminology: in texts of what domains or topics the word is usually used.

Sociolinguistics: social status, age, religion, gender, ethnicity, etc. of the speakers who typically use the word.

Correspondence to other languages: what equivalents the word has in other languages, how it can be translated.

Context in corpora: what words are used frequently/rarely with the given word, in what grammatical relations the word enter with its neighboring words, what the frequency of the word is, what their typical collocations are, etc.

4.2 Hypernym-Based Meaning Representation

In Section 3.2, we talked about automatic identification of lexical functions and considered two papers devoted to LF detection in verbal collocations, namely, (Wanner, 2004) and (Wanner *et al.*, 2006).

This section speaks of more details on how the meaning of verbal collocations was represented in the research mentioned in the previous paragraph, what meaning representation was used there in the task of assigning each verbal collocation of the structure "verb + noun" a respective lexical function.

LF identification was fulfilled by means of a few selected supervised machine learning methods which included Nearest Neighbor technique, Naïve Bayesian network, Tree-Augmented Network Classification technique and a decision tree classification technique based on the ID3-algorithm. These methods make decision of what lexical function to assign to a particular verb-noun pair on the basis of mathematical models. The models are learnt in the process of analyzing the features used in representing the meaning of verbal collocations.

The meaning of a verbal collocation in experiments by Wanner (2004) and Wanner *et al.* (2006) was represented as a union of two sets of hypernyms: the set of all hypernyms of the verb and the set of all hypernyms of the noun. The elements of a collocation, i.e., the verb and the noun, were considered as zero-level hypernyms and thus included in the respective hypernym sets. Now we will talk more about the lexical relation of hypernymy in language.

4.2.1 Hypernyms and Hyponyms

In linguistics, a **hyponym** is a word or phrase whose meaning is included within the meaning of another word, its **hypernym** (also spelled as *hypernym* in natural language processing literature). To put it simpler, a hyponym shares "A TYPE-OF" relationship with its hypernym. For example, *restaurant, rest house, planetarium, observatory, packinghouse, outbuilding, Pentagon* are all hyponyms of *building* (their hypernym), which is, in turn, a hyponym of *construction*.

In computer science, the lexical relation of hypernymy is often termed an "IS-A" relationship. For example, the phrase *Restaurant is a building* can be used to describe the hypernym relation between *restaurant* and *building*.

Thus, **hypernymy** is the semantic relation in which one word is the hypernym of another one (its hyponym).

4.2.2 WordNet as a Source of Hypernyms

WordNet (Miller, 1998; Princeton University, 2010) is an electronic lexical database where the basic means of linguistic meaning description is lexical relations. Surely, hypernymy is the most fundamental relation, perhaps, after synonymy. The importance of hypernymy is so big that for many years it was believed that the hypernymy and synonymy relations together were the necessary and sufficient features to unambiguously specify the meaning of every word. Unfortunately, it was demonstrated in practice that it is not true, so glosses (word definitions, like in conventional dictionaries) were added to complete the meaning representation of words.

- S: (n) **key#1** (metal device shaped in such a way that when it is inserted into the appropriate lock the lock's mechanism can be rotated)
    - *direct hyponym* / *full hyponym*
    - *part meronym*
    - *direct hypernym* / *inherited hypernym* / *sister term*
        - S: (n) device#1 (an instrumentality invented for a particular purpose) *"the device is small enough to wear on your wrist"*; *"a device intended to conserve water"*
            - S: (n) instrumentality#3, instrumentation#1 (an artifact (or system of artifacts) that is instrumental in accomplishing some end)
                - S: (n) artifact#1, artefact#1 (a man-made object taken as a whole)
                    - S: (n) whole#2, unit#6 (an assemblage of parts that is regarded as a single entity) *"how big is that part compared to the whole?"*; *"the team is a unit"*
                        - S: (n) object#1, physical object#1 (a tangible and visible entity; an entity that can cast a shadow) *"it was full of rackets, balls and other objects"*
                            - S: (n) physical entity#1 (an entity that has physical existence)
                                - S: (n) entity#1 (that which is perceived or known or inferred to have its own distinct existence (living or nonliving))

Fig. 2 WordNet 3.1, hypernyms for key#1.

In order to retrieve hypernyms for verbs and nouns in collocations, one must disambiguate them, since a polysemic word may have different hypernyms depending on a particular meaning. In WordNet, meanings of polysemic words are indicated by numbers.

For example, *key*#1 (metal device shaped in such a way that when it is inserted into the appropriate lock the lock's mechanism can be rotated) has hypernyms device#1; instrumentality#3, instrumentation#1; artifact#1, artefact#1; whole#2, unit#6; object#1, physical object#1; physical entity#1; entity#1 (Fig. 2).

However, *key*#2 (something crucial for explaining) has other hypernyms: explanation#2; thinking#1, thought#2, thought process#1, cerebration#1, intellection#1, mentation#1; higher cognitive process#1; process#2, cognitive process#1, mental process#1, operation#9, cognitive operation#1; cognition#1, knowledge#1, noesis#1; psychological feature#1; abstraction#6, abstract entity#1; entity#1 (Fig. 3).

- S: (n) **key#2** (something crucial for explaining) *"the key to development is economic integration"*
    - ○ *direct hypernym / **inherited hypernym** / sister term*
        - S: (n) explanation#2 (thought that makes something comprehensible)
            - S: (n) thinking#1, thought#2, thought process#1, cerebration#1, intellection#1, mentation#1 (the process of using your mind to consider something carefully) *"thinking always made him frown", "she paused for thought"*
                - S: (n) higher cognitive process#1 (cognitive processes that presuppose the availability of knowledge and put it to use)
                    - S: (n) process#2, cognitive process#1, mental process#1, operation#9, cognitive operation#1 ((psychology) the performance of some composite cognitive activity; an operation that affects mental contents) *"the process of thinking", "the cognitive operation of remembering"*
                        - S: (n) cognition#1, knowledge#1, noesis#1 (the psychological result of perception and learning and reasoning)
                            - S: (n) psychological feature#1 (a feature of the mental life of a living organism)
                                - S: (n) abstraction#6, abstract entity#1 (a general concept formed by extracting common features from specific examples)
                                    - S: (n) entity#1 (that which is perceived or known or inferred to have its own distinct existence (living or nonliving))

Fig. 3 WordNet 3.1, hypernyms for key#2.

4.2.3 An Example of Hypernym-Based Meaning Representation

As an example, let us construct a hypernym-based meaning representation of the verbal collocation *to give a smile*.

First, we have to disambiguate both words, *give* (verb) and *smile* (noun). This is not an easy task, not only for computer, but also for human annotator. The difficulty of disambiguation decision making process increases with the number of senses among which one has to choose only one sense for the context she deals with. If the reader examines the respective senses for *give* and *smile* and does not agree with us on the senses we chose, it will not be surprising at all. Inter-annotator agreement is a problem in its own right in the task of word sense disambiguation; e.g., see (Passonneau *et al.*, 2010). Clearly, the precision of sense selection has a direct correlation with the precision of machine learning methods performance on automatic detection of LF meaning, because the resulting models are obtained on the basis of choices we previously made. One possible way to resolve this problem might be experimenting with feature selection to verify what hypernyms distinguish best between lexical functions.

After considering disambiguation details, we choose the following senses for *give* and *smile*:

(v) **give#7**, throw#5 (convey or communicate; of a smile, a look, a physical gesture) *"Throw a glance"; "She gave me a dirty look"*

(n) **smile#1**, smiling#1, grin#1, grinning#1 (a facial expression characterized by turning up the corners of the mouth; usually shows pleasure or amusement)

Our case of *to give a smile* turned out to be easy because *smile* as a noun has only one sense, and *give#7* includes the noun *smile* in the list of its collocates. The simplicity of this case is just pure luck: to find *smile* listed as one of only three examples in the WordNet entry while *to give* is able to collocate with hundreds of nouns!

Now we build hypernym sets for *give* and *smile* including all their respective hypernyms together with *give#7*, *smile#1* and members of their synsets. We get the following lists:

- **hypernym set for give#7:** give#7, throw#5; communicate#2, intercommunicate#2; interact#1; act#1, move#8;
- **hypernym set for smile#1:** smile#1, smiling#1, grin#1, grinning#; facial expression#1, facial gesture#1; gesture#2, motion#1; visual communication#1; communication#2; abstraction#6, abstract entity#1; entity#1.

The meaning representation of *to give a smile* is the union of two sets of hypernyms:

to give a smile = {give#7, throw#5; communicate#2, intercommunicate#2; interact#1; act#1, move#8; smile#1, smiling#1, grin#1, grinning#; facial expression#1, facial gesture#1; gesture#2, motion#1; visual communication#1; communication#2; abstraction#6, abstract entity#1; entity#1}

In (Wanner, 2004; Wanner *et al.*, 2006), meaning representation also includes Base Concepts (BC) and Top Concepts (TC) of all hypernyms in the set exemplified in the previous paragraph for *to give a smile*. BCs and TCs were retrieved from the Spanish part of the EuroWordNet (Vossen 1998).

Base Concepts are general semantic labels or tags of semantic fields that cover a large number of synsets (recall that all words in WordNets are organized in synsets, see Section 1.3). Examples of such labels are: **change, feeling, motion**, and **possession**. The set of BCs in EuroWordNet includes 1,310 different tokens.

Each Base Concept is described in terms of Top Concepts, language-independent features such as **Agentive, Dynamic, Existence, Mental, Location, Social**, etc. (in total, 63 different TCs are distinguished in EuroWordNet).

Wanner (2004) gives an example for the Spanish verbal collocation *prestar una declaración*, to present a declaration. The example includes parts of hypernym hierarchies for *prestar* and *declaración* and indicates BSc and TCs for each hypernym synset.

4.3 Semantic Annotations as Meaning Representation

One of the research papers on automatic identification of lexical functions referenced in Section 3.2 is (Alonso Ramos *et al.*, 2008). The task was to extract pairs "support verb + object" from FrameNet corpus of examples (Ruppenhofer *et al.*, 2006) and verify if they are of the type **Oper$_n$**.

The meaning of **Oper** is 'do', 'perform', 'carry out'. The proposed algorithm made decisions on whether a particular word pair was of the meaning corresponding to LF **Oper** on the basis of several features. In this work, the features used to represent the **Oper** meaning were syntactic, semantic and collocation annotations developed for the FrameNet corpus.

Words in FrameNet are annotated with the following information:

- **Frame elements (FEs):** components of language conceptual structures or semantic frames. For example, the Commercial Transaction semantic frame consists of the following elements: possession (goods), change of possession (giving, receiving), exchange (buyer, seller), and money (price), etc. The number of elements depends on how many details needs elaboration. Another example is the Speech Communication Frame with the elements: sender, receiver, message, medium, etc. Frame elements correspond to **semantic roles**, although the first is much more detailed than the second.
- **Phrase types:** Noun Phrase (non-referential, possessive, non-maximal, snandard), Verb Phrase (with finite/non-finite verbs), Adjective Phrase, Adverb Phrase, Complement Clause, Subordinate Clause, etc.
- **Grammatical functions:** describe the ways in which constituents in a sentence satisfy abstract grammatical requirements of the target word. For example, grammatical functions assigned by verbs are external argument, object, dependent.

FrameNet data is available at https://framenet.icsi.berkeley.edu/fndrupal/home. Fig. 4 shows how the annotated text is displayed.

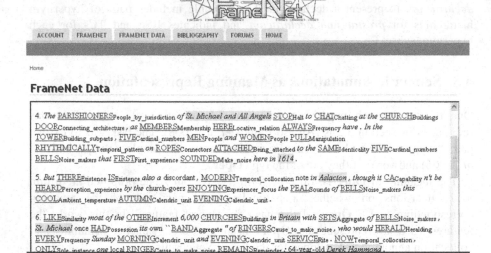

Fig. 4 FrameNet on-line user interface.

To build a meaning representation, one should choose those annotations that correspond to that meaning. Alonso Ramos *et al.* (2008) utilizes the following strategy for retrieving verb-noun pairs of $Oper_n$. First, the verb tagged as **Supp** (support) is taken to be the support verb of **Oper** (followed by any subscript), whose base is tagged as **Tgt** (target, it is a word that trigger the relations between other words, or arguments of targets). The next step is to choose tags indicative of the **Oper** subscripts, i.e., to identify what semantic tags pointed to $Oper_1$, what annotation signaled $Oper_2$, $Oper_3$, etc. The following two-part heuristic is used for the choice of **Oper** subscript:

1. If the frame element of the subject of the **Supp** is Agent, Person, Speaker, or Helper, it is assumed that this FE is the first argument and thus the subject of the underlying verb, and "1" is chosen as subscript.
2. If the FE of the subject of the **Supp** is something else, the ordering of the core FEs given in the frame of the base noun is used as a heuristic for the syntactic subcategorization. Thus, it is verified which frame element is the subject, and then the position of that FE in the list of core Fes is chosen as subscript for the **Oper**.

Consider the following two annotated utterances from FrameNet. The first utterance illustrates $Oper_1$, and the second one, $Oper_2$.

- *With reluctance, Morton decided that* [FE= AGENT *he*] *must* [Supp *make*] *another* [Tgt *attempt*] *to identify the dead girl* (Agent is subject of Supp; therefore the hypothesized LF is $Oper_1$).

- *Yet the* [FE= EVALUEE *Franks*] [Supp *have received*] [Tgt *criticism*] [Reason *for including a lot of songs dedicated fans will already own*] (Evaluee (= the second FE) is subject of Supp; therefore, the hypothesized LF is **Oper₂**).

As it was mentioned in Section 3.2, this method of meaning representation demonstrated the accuracy of 76%. Therefore, it can be concluded that extraction and semantic classification of collocations is feasible with semantically annotated corpora.

Lexical function is a concept that formalizes the correspondence between semantic roles and syntactic valency. The set of semantic roles and corresponding valency fillers in natural language utterances depends on an individual functions. Future experiments will show what existing annotations are able to function as indices of particular lexical functions.

Chapter 5
Analysis of Verbal Collocations with Lexical Functions

Verbal collocations comprise a vast group of any natural language lexis. It is important for linguistic computer applications to be capable to analyze these word combinations correctly. Indeed, the verb is the nucleus of utterance, it is the organizer of the text structure, and moreover, it is what makes a set of heterogeneous words become a message otherwise lacking its communicative force.

To analyze verbal collocations means to understand their meaning. Since collocational meaning can be so neatly and finely represented by lexical functions as we showed in Chapters 1 and 2, the task of analyzing collocations semantically can be viewed as the task of assigning a verbal collocation the lexical function which best conveys its meaning and reflects its structure.

We have just pointed out that verbal collocations are a very big group of vocabulary. For our research on verbal collocation analysis, we chose collocations of the structure "verb + complement" (*to give a comment, to launch a project, to pass an exam*) and "noun + verb" (*time passes, influence grows, trust exists*). The experiments were conducted on Spanish material. Now, as we look only at collocations with fixed syntactic structure leaving aside the issue of syntactic study, or parsing, the analysis turns out to be purely semantic.

First, we select only those lexical functions that involve the patterns "verb + complement", "noun + verb", and then we have to assign each verbal collocation in our data its lexical function.

In this chapter, we give definitions and examples of basic lexical functions corresponding to collocations of the pre-determined structure as specified in the previous paragraph; describe our data and methods of LF automatic detection, and present results of our experiments.

5.1 Verbal Lexical Functions

5.1.1 Basic Lexical Functions in Verbal Collocations

Here we will explain the meaning of most common lexical functions in collocations following the structural patterns "verb + complement", "noun + verb" and give examples.

A. Gelbukh, O. Kolesnikova: Semantic Analysis of Verbal Collocations, SCI 414, pp. 61–83.
springerlink.com   © Springer-Verlag Berlin Heidelberg 2013

Before presenting verbal LFs, we need to comment on a few concepts involved in semantic analysis of verbal collocations. Recall that collocations consist of two fundamental elements: the base and the collocate. The **base** is the word used in its principal and typical meaning. The distinctive feature of the **collocate** is that in collocations such word is used to express a meaning different from its typical meaning conveyed by it in free word combinations. To illustrate this peculiarity of collocation let us consider an example.

One of most common collocations is *to make a decision*. Here we will remark that it is a lexicographic convention that the most frequent and therefore typical sense is given the highest rank in the list of senses and is put first in the dictionary entry. The first sense of *to make* in WordNet 3.1 online (Princeton University, 2010) is the following:

(v) **make#1**, do#1 (engage in) *"make love, not war"; "make an effort"; "do research"; "do nothing"; "make revolution"*.

The sense *to make* has in *to make a decision* is given the 16th rank:

(v) **make#16** (perform or carry out) *"make a decision"; "make a move"; "make advances"; "make a phone call"*.

At the same time, the meaning *a decision* has in *to make a decision* is the most frequent one as of WordNet 3.1:

(n) **decision#1**, determination#5, conclusion#9 (the act of making up your mind about something) *"the burden of decision was his"; "he drew his conclusions quickly"*.

Collocations differ from true idioms (*a frog in the throat, kill some time, kick the bucket, spill the beans*) in the sense that idioms are fully non-compositional expressions, but the semantic of collocations is to some extent compositional. It is revealed in the fact that bases in collocations preserve their typical meanings and thus can be interpreted literally. The disambiguation task is associated therefore with collocates and lexical functions is a perfect tool to fulfill such task.

It has been observed by linguists that collocations are typical for journalistic, business, academic and technical writing and most noun bases in verbal collocations denote actions, processes, states, i.e., the same type of semantics that verbs also possess, for that reason such nouns are called **deverbal nouns**.

The verb is characterized by its **subcategorization frame**, or other words called **syntactic arguments**. In *The cat sat on a mat* the verb *sit* has two arguments, *the cat* and *the mat*.

Now let us speak not in terms of syntax and syntactic arguments, but in terms of meaning and what happens in real life. In this world, there exists a situation or state of affairs lexicalized as *sit*. The *sit*-situation is not "all-in-one-piece" but a system of "components" including, first of all, the **action** of sitting itself, secondly, the one who performs the action of sitting (**agent**) and lastly, the **object** on which the agent sits. We will say that the action *sit* has two arguments, agent and object. In another terminological tradition, it is conventional to say that *sit* has two **semantic actants** – agent and object.

At this point we will go back to syntax and observe what words are used to denote or lexicalize the agent and the object. In other words, what **syntactic actants** "play the role" of respective semantic actants. So speaking in terms of syntactic actants or **syntactic functions**, in *The cat sat on the mat*, the first semantic actant (agent) is expressed by the syntactic actant called the subject of utterance (*the cat*), and the second semantic actant is expressed by indirect object (*the mat*).

Syntagmatic verbal LFs generalize the meaning of collocates into several semantic classes whose meaning is reflected by the LF name. Lexical functions are named by abbreviated Latin words with the meaning closest to the semantics of a given class. At the same time, LFs encode information on semantic and syntactic actants of collocational bases.

Here we present some basic verbal LFs, specify their meaning and give examples chosen by us as well as borrowed from (Mel'čuk, 1996; Wanner, 2004; Wanner *et al.*, 2006). The following abbreviations are used in examples: sth (something), sb (somebody).

Oper$_1$ (*operare*, to do, perform, carry out)– a light (support) verb which connects the first semantic actant (agent) lexicalized as subject with the name of the situation (keyword) lexicalized as the direct object: *to take a bath/breath/ vacation, to have a look/sleep/talk/holiday/bath/shower, to give* sb/sth *a smile/laugh/shout/push, to conduct a survey/experiment/inquiry, to carry out repairs/a plan, to ask a question, to teach a lesson.*

Oper$_2$ – a light verb which connects the second semantic actant (patient, recipient, experiencer) lexicalized as subject with the keyword lexicalized as direct object: *to get a benefit, to have an attack (of a disease), to receive treatment, to gain attention, to sit an exam, to face an accusation, to undergo an inspection, to take advice.*

Func$_0$ (*functionare*, to function) – a light verb meaning 'to take place' which syntactically is the predicate of the keyword lexicalized as subject: *the possibility exists, time flies, the day passes by, a doubt arises, joy fills* (sb), *the wind blows, an accident happens, the rain falls.*

Func$_1$ – a light verb which connects the keyword as subject with the first semantic actant (agent) as object: *responsibility lies* (with sb), *the analysis is due* (to sth), *the blow comes* (from sb), *the proposal stems* (from sth), *support comes* (from sb).

Func$_2$ – a light verb which connects the keyword as subject with the second semantic actant (agent) as object: *the analysis concerns* (sth), *the blow falls* (upon sb), *the change affects* (sth), *the lecture is* (on the theme of sth).

Labor$_{ij}$ (*laborare*, to work, toil) – a light verb which connects the ith semantic actant as subject with the jth semantic actant as direct object (principal object) and the keyword as a secondary object. Example: **Labor$_{12}$**(*interrogation*) = *to submit* (sb) *to an interrogation, to keep* (sb) *under control, to treat* (sb) *with respect, to subject* (sb) *to punishment*; **Labor$_{321}$**(*lease*) = *to grant* (sth to sb) *on lease.*

Copul (*copula*) – a verb whose purpose is to"verbalize" noun (keyword): *to be a teacher, to work as a teacher, to be an example, to represent an example, to serve as an example*

Manif (*manifestare*, to manifest) – *amazement lurks, joy explodes* (in sb), *scorn is dripping* (from every act).

Degrad (*degradare*, to lower, degrade) – *clothes wear off, discipline decays, the house becomes dilapidated, milk goes sour, patience wears thin, teeth decay, temper frays.*

Real$_1$ (*realis*, real) – a verb meaning 'to fulfill the requirement (of the keyword)', 'to do (with the keyword) what you are supposed to do (with it)': *to acknowledge a debt, to prove the accusation, to keep a promise, to stick to the schedule, to fulfill a threat, to drive a bus/car, to succumb to an illness.*

Real$_2$ – *to ride on a bus, to meet a demand, to take a hint, to abide by a law.*

LFs that are not used independently but only as elements of complex LFs:

Incep (*incipere*, to begin)
Cont (*continuare*, to continue)
Fin (*finire*, to cease)
Caus (*causare*, to cause)
Liqu (*liquidare*, to liquidate)
Perm (*permittere*, to permit, allow)

Examples of complex LFs:

IncepOper$_1$ – *to open fire* (on sth), *to acquire popularity, to sink into despair*
IncepOper$_2$ – *to fall under the power* (of sb), *to get under one's control*
IncepFunc$_1$ – *despair creeps over/in* (sb), *hatred stirs up* (sb), *anger arises*
ContOper$_1$ – *to retain one's power* (over sth), *to maintain the supremacy, to maintain enthusiasm*
ContOper$_2$ – *to hold attention*
ContFunc$_0$ – *the offer stands, the odor lingers*
FinOper$_1$ – *to lose one's power* (over sb)
FinOper$_2$ – *to lose credit* (with sb)
FinFunc$_0$ – *anger defuses, hatred ceases, enthusiasm disappears*
LiquFunc$_0$ – *to stop an aggression, to dissolve an assembly, to wipe out a trace*
LiquFunc$_2$ – *to divert one's attention* (from sth)
CausOper$_1$ – *to lead* (sb) *to an opinion, to reduce* (sb) *to despair, to throw* (sb) *into despair*
CausFunc$_0$ – *to bring about a crisis, to create a difficulty, to present a difficulty*
CausFunc$_1$ – *to raise hope* (in sb)
CausFunc$_1$ – *to raise hope* (in sb)
PermFunc$_0$ – *to condone an aggression*

5.1.2 Lexical Functions Chosen for our Experiments

Our choice of lexical functions depends on the number of examples that each lexical function has in the lexical resource of Spanish lexical functions created by us and described in Chapter 7. We have selected LFs that have the number of examples sufficient for machine learning experiments. Wanner (2004) and Wanner *et al.* (2006) experimented with the following number of LF examples: the biggest number of examples that this researcher had in the training set was 87 for **Oper$_1$** and the least number of examples was 33 for **Oper$_2$**. This data is characterized in Section 3.2.

Table 10 presents LFs that we have chosen for our experiments. For each LF, we give the number of examples, its meaning, and sample verb-noun combinations. In the second column "K" means keyword or lexical function argument.

Table 10 Lexical functions chosen for our experiments, K stands for the keyword.

LF and # of examples	Meaning	Collocation: LF value + keyword	
		Spanish	English translation
Oper$_1$ 280	Experience (if K is an emotion), carry out K.	*alcanzar un objetivo* *aplicar una medida* *corregir un error* *satisfacer una necesidad*	*achieve a goal* *apply a measure* *correct a mistake* *satisfy a necessity*
CausFunc$_0$ 112	Do something so that K begins occurring.	*encontrar respuesta* *establecer un sistema* *hacer campaña* *producir un efecto*	*find an answer* *establish a system* *conduct a campaign* *produce an effect*
CausFunc$_1$ 90	A person/object, different from the agent of K, does something so that K occurs and has effect on the agent of K.	*abrir camino* *causar daño* *dar respuesta* *producir un cambio*	*open the way* *cause damage* *give an answer* *produce a change*
Real$_1$ 61	To fulfill the requirement of K, to act according to K.	*contestar una pregunta* *cumplir el requisito* *solucionar un problema* *utilizar la tecnología*	*answer a question* *fulfill the requirement* *solve a problem* *use technology*
Func$_0$ 25	K exists, takes place, occurs.	*el tiempo pasa* *hace un mes* *una posibilidad cabe* *la razón existe*	*time flies* *a month ago* *there is a possibility* *the reason exists*
Oper$_2$ 30	Undergo K, be source of K	*aprender una lección* *obtener una respuesta* *recibir ayuda* *sufrir un cambio*	*learn a lesson* *get an answer* *receive help* *suffer a change*
IncepOper$_1$ 25	Begin to do, perform, experience, carry out K.	*adoptar una actitud* *cobrar importancia* *iniciar una sesión* *tomar posición*	*take an attitude* *acquire importance* *start a session* *obtain a position*
ContOper$_1$ 16	Continue to do, perform, experience, carry out K.	*guardar silencio* *mantener el equilibrio* *seguir un modelo* *llevar una vida (ocupada)*	*keep silence* *keep one's balance* *follow an example* *lead a (busy) life*

In the Dictionary of Spanish Lexical Functions, we have annotated **free word combinations** with the tag **FWC**. The number of FWCs is 261. We considered free word combinations as a lexical function FWC in its own right and experimented how machine learning algorithms can predict this class of word combinations. Therefore, the total number of LFs we experimented with is nine.

5.2 Supervised Machine Learning Algorithms

Our approach is based on supervised machine learning algorithms as implemented in the WEKA version 3-6-2 toolset (Hall *et al.*, 2009), (Witten and Frank, 2005). We evaluated the prediction of LFs meanings on the training sets of collocations from the Dictionary of Spanish Verbal Lexical Functions (Chapter 7) using 10-fold cross-validation technique. Table 11 lists all 68 machine learning algorithms we experimented with.

Table 11 Machine learning algorithms used in our experiments

Algorithm	Algorithm	Algorithm
AODE	ClassificationViaClustering	VFI
AODEsr	ClassificationViaRegression	ConjunctiveRule
BayesianLogisticRegression	CVParameterSelection	DecisionTable
BayesNet	Dagging	JRip
HNB	Decorate	NNge
NaiveBayes	END	OneR
NaiveBayesSimple	EnsembleSelection	PART
NaiveBayesUpdateable	FilteredClassifier	Prism
WAODE	Grading	Ridor
LibSVM	LogitBoost	ZeroR
Logistic	MultiBoostAB	ADTree
RBFNetwork	MultiClassClassifier	BFTree
SimpleLogistic	MultiScheme	DecisionStump
SMO	OrdinalClassClassifier	FT
VotedPerceptron	RacedIncrementalLogitBoost	Id3
Winnow	RandomCommittee	J48
IB1	RandomSubSpace	J48graft
IBk	RotationForest	LADTree
KStar	Stacking	RandomForest
LWL	StackingC	RandomTree
AdaBoostM1	ThresholdSelector	REPTree
AttributeSelectedClassifier	Vote	SimpleCart
Bagging	HyperPipes	

We evaluated the performance of the selected algorithms by comparing precision, recall, and *F-measure* (values for predicting the positive class). The precision is the proportion of the examples which truly have class x among all those which were classified as class x. The recall is the proportion of examples which were classified as class x, among all examples which truly have class x.

The *F-measure* is the harmonic mean of precision and recall:

$$F = 2 \times \frac{\text{Precision} \times \text{Recall}}{\text{Precision} + \text{Recall}}.$$

5.2.1 WEKA Data Mining Toolkit

WEKA (Waikato Environment for Knowledge Analysis) is well-known software for machine learning and data mining developed at the University of Waikato. This program is written in Java. WEKA is an open-source workbench distributed under the GNU-GPL license. For machine learning experiments, we have used WEKA version 3-6-2, see References. WEKA workbench has a graphical user interface that leads the user through data mining tasks and has data visualization tools that help understand the models.

5.2.2 Stratified 10-Fold Cross-Validation Technique

For evaluating the prediction of machine learning algorithms on the training set, we have used stratified 10-fold cross-validation technique. The simplest form of evaluating the performance of classifiers is using a training set and a test set which are mutually independent. This is referred to as hold-out estimate.

We have chosen a more elaborate evaluation method, i.e. cross-validation. Here, a number of folds n is specified. The dataset is randomly reordered and then split into n folds of equal size. In each iteration, one fold is used for testing and the other $n - 1$ folds are used for training the classifier. The test results are collected and averaged over all folds. This gives the cross-validation estimate of the accuracy. The folds can be purely random or slightly modified to create the same class distributions in each fold as in the complete dataset. In the latter case the cross-validation is called stratified. We have applied the stratified option of 10-fold cross-validation method.

5.3 Representing Data for Machine Learning Techniques

Recall that in the data set, each verb-noun combination is represented as a set of all hypernyms of the noun and all hypernyms of the verb. To construct this representation, the sense number for every verb and every noun must be identified. So we disambiguated all words with word senses of the Spanish part of EuroWordNet, or Spanish WordNet (Vossen, 1998). But sometimes, an appropriate sense was absent in the Spanish WordNet. Such words were tagged in the Dictionary (Chapter 7) with abbreviation N/A (not available) instead of the number of word sense. In the training set, we included only verb-noun combinations that are disambiguated with word senses of the Spanish WordNet. In Table 10, the numbers of examples include only these verb-noun pairs in which all the words are disambiguated with the Spanish WordNet.

The total number of examples for all nine lexical functions is 900.

5.3.1 Training (Data) Sets

For each of the 9 LFs chosen for experiments, we built a training set, so we had 9 training sets. All training sets included the same list of 900 verb-noun combinations. The only difference between training sets was the annotation of examples as positive and negative.

As an example, let us consider the training set for **Oper$_1$**. In the list of 900 verb-noun pairs, there are 266 examples of **Oper$_1$**, so these examples are marked as positive in the training set, and all the rest of verb-noun combinations whose number is 634 (900 − 266 = 634) were marked as negative examples. This procedure was applied to each training set.

In Chapter 4, we discussed meaning representation of linguistic data. We chose to apply hypernym-bease representation of lexical lada. Therefore, each verb-noun pairs was represented as a set of all hypernyms of the noun and all hypernyms of the verb.

However, the machine learning techniques can access data only if it is represented in the Attribute-Relation File Format (ARFF). A dataset in ARFF format is a collection of examples, each one of class `weka.core.Instance`. Recall that WEKA is written in Java. Each `Instance` consists of a number of attributes, any of which can be nominal (one of a predefined list of values), numeric (a real or integer number) or a string (an arbitrary long list of characters, enclosed in "double quotes"). In our case, all attributes are nominal.

For each verb-noun pair, we used binary feature representation. Every hypernym is represented as a nominal attribute which can take one of two values: "1" if it is a hypernym of any word in a given verb-noun pair, and "0" if it is not. The fact that a verb-noun combination belongs or does not belong to a particular LF is identified by the class attribute with two possible values: "yes" for positive examples of a particular LF and "no" for negative ones.

All training sets include 900 verb-noun pairs represented by hypernyms. This gives 798 features to represent the nouns, and 311 features to represent the verbs. Total number of features that are hypernyms is 1109. There is also one class attribute; therefore, each training set includes 1110 attributes. Verb-noun pairs in the training set are represented as vectors of length 1110:

$$v_1, v_2, ..., v_{798}, n_1, n_2, ..., n_{311}, LF,$$

where v_n, n_k can be 0 or 1, and *LF* is a class attribute having the value *yes* for positive instances of LF for which classification is done, and *no* for negative instances. A partial representation of the training set in the ARFF format is given in Fig. 5.

```
@relation Oper1
@attribute n00001740 {0,1} % entidad_1
@attribute n00002086 {0,1} % ser_vivo_1
@attribute n00003731 {0,1} % agente_causal_1 causa_4
...
@attribute v01128460 {0,1} % causar_5 producir_4
ocasionar_1
@attribute v01130277 {0,1} % hacer_2
@attribute v01131926 {0,1} % dar_9
...
@attribute category {yes,no}

@data
1,0,0,0,1,0,0,1,0,0,0,0,0,...,0,0,0,yes %
v_hacer_15 n_mención_1
...
0,0,0,0,0,0,0,0,0,0,0,0,0,1,...,0,0,0,no % v_abrir_5
n_camino_5
```

Fig. 5 A partial representation of the training set in the ARFF format.

In Fig. 5, after the tag `@relation`, we define the internal name of the dataset. Nominal attributes are defined with the tag `@attribute`. The record `@data` marks the beginning of the data section. The data section consists of comma-separated values for the attributes – one line per example. In Fig. 5, we show only two examples: one positive and one negative. All records beginning with `@attribute` as well as all strings in the data section are accompanied by a comment starting with symbol % in ARFF. For attribute entries, comments include words of the synset specified by its number after the record `@attribute`. For data strings, comment includes the verb-noun pair represented by this string. The pair is annotated with POS (part-of-speech) and word senses. Comments were added for human viewing of data.

The process of training set compilation explained above can be represented as fulfilled according to the algorithm shown in Fig. 6.

Algorithm: constructing data sets
Input: a list of 900 Spanish verb-noun collocations annotated with 8 lexical functions
Output: 8 data sets – one for each lexical function
For each lexical function
Create an empty data set and assign it the name of the lexical function.
For each collocation in the list of verb-noun collocations
        Retrieve all hypernyms of the noun.
        Retrieve all hypernyms of the verb.
        Make a set of hypernyms:
        {noun, all hypernyms of the noun, verb, all hypernyms of the verb}.
        If a given collocation belongs to this lexical function
            assign "1" to the set of hypernyms,
        Else assign "0" to the set of hypernyms.
        Add the set of hypernyms to the data set.
Return the data set.

Fig. 6 Algorithm of compiling the training sets.

5.4 Classification Procedure

Fig. 7 presents the classification procedure schematically. Since we used 10-fold cross-validation technique (described in Section 5.2) to evaluate the performance of algorithms, we speak only of the training set, or data set, without mentioning a test set. However, in Fig. 7 both sets are shown. The matter is that during the execution of learning algorithms, the data set is iteratively split into a proper training set and a test set according to the procedure of 10-fold cross validation.

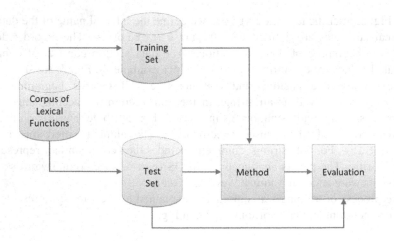

Fig. 7 The classification procedure.

Now, we will explain the classification procedure using the WEKA graphical user interface. First, the training set is downloaded and WEKA shows various characteristics of the data in Explorer, a part of the interface developed for the purpose of analyzing data in the data set. This stage is demonstrated in Fig. 8.

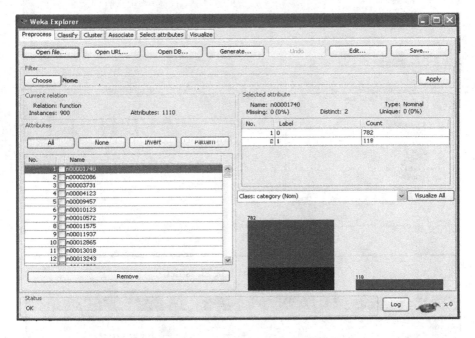

Fig. 8 Characteristics of data viewed in WEKA Explorer.

Secondly, a classifier is chosen for data classification. This step can be vied in Fig. 9.

Thirdly, the chosen classifier starts working and after finishing the learning and the test stages, it outputs various estimates of its performance. In our experiments, we used the values of precision, recall and F-measure for "yes" class, or the positive class, to evaluate the performance of the classifiers. A classifier's output is demonstrated in Fig. 10.

While for initial experiments the included graphical user interface is quite sufficient, for in-depth usage the command line interface is recommended, because it offers some functionality which is not available via the graphical user interface (GUI). When data is classified with the help of GUI, the default heap memory size is 16–64 MB. For our data sets, this is too little memory. The heap size of java engine can be increased via −Xmx1024m for 1GB in the command line. Also, via the command line using the −cp option we set CLASSPATH so that includes weka.jar.

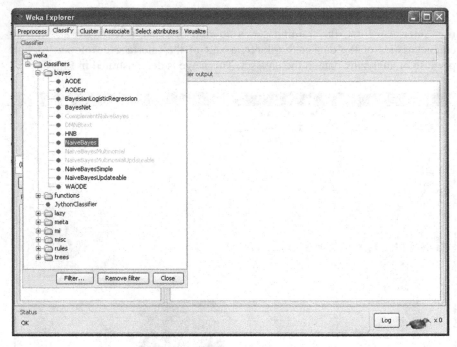

Fig. 9 Selection of the classifier.

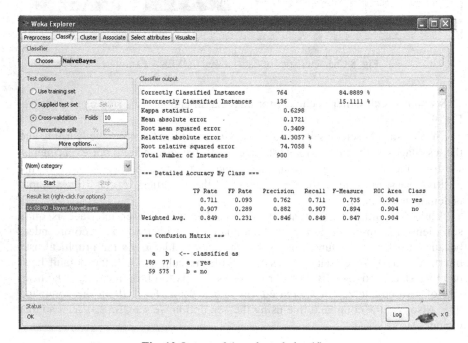

Fig. 10 Output of the selected classifier.

Taking the above advantages into account, we operated WEKA through the command line using the following options for the algorithm PART of class rules taken as an example:

```
set classpath=%classpath%;C:\Archivos de
programa\Weka-3-6\weka.jar

set classpath=%classpath%;C:\Archivos de
programa\Weka-3-6\rules.jar

set classpath=%classpath%;C:\Archivos de
programa\Weka-3-6\PART.jar

java -Xmx1024m weka.classifiers.rules.PART -t
training_set.arff -d part.model

java  Xmx1024m weka.classifiers.rules.PART -l
part.model -T test_set.arff -p 0 >> result1.txt 2>
$$1.tmp
```

The first three lines of the above example show how CLASSPATH is set for WEKA and the chosen classifier in particular. The fourth line tells the classifier to learn on the training set and to save the model created for the training data in a file with the extension model. The last line commands the classifier to test the learnt model on the test set and save the results of prediction in a file. To preserve error messages of the classifier for further debug, the standard error stream (2>) can be directed to a file and saved in it.

5.5 Algorithm Performance Evaluation

In Table 12, we present the main results obtained in our experiments. For each LF, we list three machine learning algorithms that have shown the best performance. The symbol # stands for the number of examples in the respective training set; P stands for precision, R stands for recall, F stands for *F-measure*.

Results of all algorithms evaluated in our experiments are given in Tables 1–9 of Appendix.

5.5.1 *Baseline*

Often, in classification experiments, the baseline is the performance of ZeroR classifier. ZeroR is a trivial algorithm that always predicts the majority class. It happens that the majority class in our training sets is always the class of negative instances. Even in the case of the LF which has the largest number of positive instances in the training set (280 positive examples of $\mathbf{Oper_1}$), the number of negative instances is still larger ($900 - 280 = 620$ negative examples of $\mathbf{Oper_1}$). Therefore, the ZeroR does not classify any test instances as positives, which gives always recall of 0 and undefined precision. Thus ZeroR is too bad a baseline to be considered.

Tabel 12 Best results shown by algorithms on the training set of lexical functions

LF	#	Algorithm	P	R	F	Baseline
$Oper_1$	280	BayesianLogisticRegression	0.879	0.866	**0.873**	
		Id3	0.879	0.861	0.870	0.311
		SMO	0.862	0.866	0.864	
$CausFunc_0$	112	JRip	0.747	0.705	**0.725**	
		EnsembleSelection	0.744	0.659	0.699	0.124
		REPTree	0.750	0.648	0.695	
$CausFunc_1$	90	J48	0.842	0.696	**0.762**	
		FilteredClassifier	0.842	0.696	0.762	0.100
		END	0.842	0.696	0.762	
$Real_1$	61	Prism	0.735	0.832	**0.781**	
		BayesianLogisticRegression	0.788	0.553	0.650	0.068
		SMO	0.722	0.553	0.627	
$Func_0$	25	BFTree	0.667	0.727	**0.696**	
		Id3	0.571	0.727	0.640	0.028
		AttributeSelectedClassifier	0.636	0.636	0.636	
$Oper_2$	30	PART	0.923	0.571	**0.706**	
		AttributeSelectedClassifier	0.923	0.571	0.706	0.033
		END	0.923	0.571	0.706	
$IncepOper_1$	25	Prism	0.750	0.800	**0.774**	
		NNge	0.923	0.600	0.727	0.028
		SMO	0.813	0.650	0.722	
$ContOper_1$	16	SimpleLogistic	0.909	0.769	**0.833**	
		DecisionTable	0.909	0.769	0.833	0.018
		AttributeSelectedClassifier	0.833	0.769	0.800	
FWC	261	Prism	0.639	0.702	**0.669**	
		BayesianLogisticRegression	0.658	0.629	0.643	0.290
		SMO	0.656	0.623	0.639	
Total:	900			*Average best:*	0.758	

However, the baseline can be a random choice of a positive or a negative answer to the question "Is this collocation of this particular lexical function?" In such a case we deal with the probability of a positive and negative response.

Since we are interested in only assigning the positive answer to a collocation, we calculate the probability of "yes" class for eight lexical functions in the experiments according to the formula: probability of "yes" = 1 / (the number of all examples / the number of positive examples of a given lexical function). These probabilities will be results of a classifier that assigns the class "yes" to collocations at random. Since we will compare the probabilities of the random choice with the results obtained in our experiments, we present the former as numbers within the range from 0 to 1 in Table 13 as well as in Table 12.

Table 13 Probability of selecting "yes" class at random

Lexical function	Number of examples	Probability of the class "yes"
$Oper_1$	280	0.311
$CausFunc_0$	112	0.124
$CausFunc_1$	90	0.100
$Real_1$	61	0.068
$Func_0$	25	0.028
$Oper_2$	30	0.033
$IncepOper_1$	25	0.028
$ContOper_1$	16	0.018
FWC	261	0.290

5.5.2 Three Best Machine Learning Algorithms for Each Lexical Function

As it is seen from Table 12, no single classifier is the best one for detecting all LFs. For each LF, the highest result is achieved by a different classifier. However, Prism reaches the highest F-score for both $IncepOper_1$ and FWC, though recall that FWC (free word combinations) is not a lexical function but is considered as an independent class along with LFs. The maximum F-measure of 0.873 is achieved by BayesianLogisticRegression classifier for $Oper_1$. The lowest best F-measure of 0.669 is shown by Prism for FWC. The average F-measure (calculated over only the nine best results, one for each LF) is 0.758.

We observed no correlation between the number of instances in the training set and the results obtained from the classifiers. For example, a low result is shown for the class FWC which has the largest number of positive examples. On the contrary, the second top result is achieved for LF $ContOper_1$, with the smallest number of positive examples. The minimum F-measure is obtained for FWC whose number of positive examples (261) is a little less than the largest number of positive examples ($Oper_1$ with 280 examples) but the detection of $Oper_1$ was the best.

For comparison, Table 14 gives the state of the art results reported in (Wanner et al., 2006) for LF classification using machine learning techniques. Out of nine LFs mentioned in (Wanner et al., 2006) we give in Table 14 only those five that we used in our experiments, i.e., that are represented in Table 12. The numbers of our results are rounded to include two figures after the point, since the results of (Wanner et al., 2006) are represented in this manner. Also, as we have explained in Section 3.2 that Wanner et al. (2006) reports the results for two different datasets: one for a narrow semantic field (that of emotions) and another for a field-independent (general) dataset. Since our dataset is also general, comparing them with a narrow-field dataset would not be fair, so in Table 14 we only give the field-independent figures from (Wanner et al., 2006).

Table 14 State of the art results for some LFs taken from (Wanner *et al.*, 2006)

LF	NN			NB			ID3			TAN			Our
	P	R	F	P	R	F	P	R	F	P	R	F	F
$Oper_1$	0.65	0.55	0.60	0.87	0.64	**0.74**	0.52	0.51	0.51	0.75	0.49	0.59	**0.87**
$Oper_2$	0.62	0.71	**0.66**	0.55	0.21	0.30	N/A			0.55	0.56	0.55	**0.71**
$ContOper_1$	N/A			N/A			0.84	0.57	**0.70**	N/A			**0.83**
$CausFunc_0$	0.59	0.79	**0.68**	0.44	0.89	0.59	N/A			0.45	0.57	0.50	**0.73**
$Real_1$	0.58	0.44	**0.50**	0.58	0.37	0.45	N/A			0.78	0.36	0.49	**0.78**
Best average: 0.66											*Average:*		0.78

Not all methods have been applied in (Wanner *et al.*, 2006) for all LFs; if a method was not applied for a particular LF, the corresponding cells are marked as N/A. In this table, NN stands for the Nearest Neighbor technique, NB for Naïve Bayesian network, ID3 is a decision tree classification technique based on the ID3-algorithm, and TAN for the Tree-Augmented Network Classification technique; P, R, and F are as in Table 12. In fact, Wanner *et al.* (2006) did not give the value of F-measure, so we calculated it using the formula in Section 5.2. The last column repeats the best F-measure results from Table 12, for convenience of the reader. For each LF, the best result from (Wanner *et al.*, 2006), as well as the overall best result (including our experiments), are marked in boldface.

As seen from Table 14, for all LFs our experiments gave significantly higher figures than those reported in (Wanner *et al.*, 2006). The best average F-measure from (Wanner *et al.*, 2006) is 0.66, while our experiments demonstrate the best average F-measure of 0.75 (calculated from Table 12) and the average F-measure is 0.78.

However, the comparison is not fair because different datasets have been used: the exact dataset used in (Wanner *et al.* 2006) is unfortunately not available anymore, ours is available from www.Gelbukh.com/lexical-functions.

5.5.3 Algorithm Performance on the Training Set

In the tables 1–9 of Appendix, we present the results of performance of 68 machine learning algorithms on 9 training sets, i.e., one training set for each of 9 LFs chosen for the experiments. As in previous tables, P stands for precision, R stands for recall and F stand for F-measure. All algorithms are ranked by F-measure.

5.5.4 Three Hypothesis in (Wanner et al., 2006) and our Results

In Section 3.2, we mentioned three hypotheses expressed in (Wanner *et al.*, 2006). These hypotheses are three possible methods of how humans recognize and learn collocations. Now we formulate these three hypothetic methods and comment them using our results.

Method 1. Collocations can be recognized by their similarity to the prototypical sample of each collocational type; this was modeled by Nearest Neighbor technique. WEKA implements the nearest neighbor method in the following classifiers: NNge, IB1, IBk and KStar (Witten and Frank, 2005).

NNge belongs to the class `rules`; classifiers of this type are effective on large data sets, and our training sets include many examples (900) and possess high dimensionality. NNge is an algorithm that generalizes examples without nesting or overlap.

NNge is an extension of Nge, which performs generalization by merging examples exemplars, forming hyper-rectangles in feature space that represent conjunctive rules with internal disjunction. NNge forms a generalization each time a new example is added to the database, by joining it to its nearest neighbor of the same class.

Unlike Nge, it does not allow hyper-rectangles to nest or overlap. This is prevented by testing each prospective new generalization to ensure that it does not cover any negative examples, and by modifying any generalizations that are later found to do so. NNge adopts a heuristic that performs this post-processing in a uniform fashion. The more details about this algorithm can be found in (Roy, 2002).

In spite of its advantages, NNge does no perform very well on predicting LFs. The only LF which NNge predicts well is **IncepOper$_1$** with the meaning 'to begin doing something' (F-measure of 0.727). For the rest eight LFs NNge does not show high results: for **Oper$_1$** the value of F-measure is 0.817, for **ContOper$_1$** it is 0.750, for **CausFunc$_1$** the value of F-measure is 0.702, for **FWC** it is 0.606, for **Oper$_2$** it is 0.600, for **Real$_1$** it is 0.593, for **CausFunc$_0$** it is 0.543, for **Func$_0$** it is 0.522. The average F-measure for all nine LFs is 0.651.

Classifiers IB1, IBk and KStar, also based on the nearest neighbor algorithm, belong to the class lazy. It means that their learning time is very short, in fact learning in its full sense does not take place in these algorithms. An unseen example is compared with all instances annotated with LFs, and the LF whose instance is closer to the unseen example, is assigned to the latter.

On our training sets, IB1, IBk and KStar show worse performance than NNge. Here we give average values of F-measure for IB1, IBk and KStar. The average value of F-measure for **Oper$_1$** is 0.616, for **FWC** it is 0.589, for **CausFunc$_1$** it is 0.466, for **ContOper$_1$** it is 0.410, for **Oper$_2$** it is 0.407, for **CausFunc$_0$** it is 0.404, for **IncepOper$_1$** it is 0.395, for **Real$_1$** it is 0.378, for **Func$_0$** it is 0.340. The average F-measure for all nine LFs is 0.445. It means that it is difficult to find a very distinctive prototypical LF instance that could indeed distinguish the meaning of a particular LF.

IncepOper$_1$ is an exception here, and the distance between the examples of IncepOper$_1$ and the examples of all the rest of LF is significantly bigger that the distance between the examples of **IncepOper$_1$**. But for eight LFs, our results demonstrate that Method 1 does not produce high quality results.

Method 2. Collocations can be recognized by similarity of semantic features of collocational elements to semantic features of elements of collocations known to belong to a specific LF; this was modeled by Naïve Bayesian network and a decision tree classification technique based on the ID3-algorithm.

We tested three WEKA Naïve Bayesian classifiers: NaiveBayes, NaiveBayesSimple, and NaiveBayesUpdateable (Witten and Frank, 2005). All three classifiers show low results for $Oper_1$ (F-measure of 0.711), **FWC** (F-measure of 0.546), $CausFunc_0$ (F-measure of 0.308), $CausFunc_1$ (F-measure of 0.077), $Real_1$ (F-measure of 0.040).

These three Bayesian classifiers failed to predict $Func_0$, $Oper_2$, $IncepOper_1$, $ContOper_1$; their result for these LFs is 0.

Bayesian classifiers are based on the assumption that attributes used to represent data are independent. As the results show, this is not the case with our training sets. Indeed, the features used to represent verb-noun combinations are hypernyms of the verb and hypernyms of the noun. Hypernyms are organized in a hierarchy, and are placed in a fixed order in the branches of their taxonomy, so that the position of each hypernym depends on the location of other hypernyms in the same branch.

Moreover, it seems that in verb-noun pairs, the verb depends on the noun since they both form a collocation but not a free word combination where there is no lexical dependency between the constituents. These are the reasons why Bayesian algorithms perform poorly on LFs. Since Bayesian methods intent to model human recognition of collocational relations, it can be concluded, that our results for Bayesian classifiers do not support the hypothesis of learning collocations by similarity of semantic features of collocational elements to semantic features of elements of collocations known to belong to a specific LF.

Now, let is consider another machine learning algorithm that models the same method of collocation recognition, ID3 algorithm. This algorithm is implemented in WEKA as Id3 classifier of the class `trees`.

Id3 classifier shows the top result for $CausFunc_1$ as compared with the performance of the other classifiers for predicting the same LF, (F-measure of 0.762), the second best result for predicting $Oper_1$ (F-measure of 0.870), another second best result for $ContOper_1$ (F-measure of 0.800). For $Oper_2$ the value of F-measure is 0.706, for $IncepOper_1$ it is 0.667, for $Func_1$ the value of F-measure is 0.636, for **FWC** it is 0.631, for $CausFunc_0$ the value of F-measure is 0.621, for $Real_1$ it is 0.587.

The average F-measure for all nine LFs is 0.698. This classifier gives average results close to the results of NNge of Method 1 but which still are rather low. However, for $Oper_1$ and $ContOper_1$ the results are quite satisfactory. It means that the semantic features of these two lexical function which in our case are hypernyms, distinguish sufficiently well these two lexical functions from the rest seven LFs.

Method 3. The third method was modeled by Tree-Augmented Network (TAN) Classification technique. As it is seen from results in (Wanner *et al.*, 2006) demonstrated in Table 5, the nearest neighbor algorithm gives better results in terms of recall than TAN. So it can be concluded that there are more evidence in favor of Method 2 than Method 3. We did not apply TAN method in our experiments.

5.5.5 Best Machine Learning Algorithms

Now we consider the best algorithm for each LF chosen for our experiments.

Oper₁: the Best Algorithm is BayesianLogisticRegression

For **Oper₁**, BayesianLogisticRegression has shown precision of 0.879, recall of 0.866, and F-measure of 0.873.

Logistic Regression is an approach to learning functions of the form $f : X \rightarrow Y$, or $P(Y|X)$ in the case where Y is discrete-valued, and $X = \{X_1 ...X_n\}$ is any vector containing discrete or continuous variables. In the case of our training data, the variables are discrete. X is a variable for the attributes, and Y is a Boolean variable which corresponds to the class attribute in the training set.

Logistic Regression assumes a parametric form for the distribution $P(Y|X)$, then directly estimates its parameters from the training data. Logistic Regression is trained to choose parameter values that maximize the conditional data likelihood. The conditional data likelihood is the probability of the observed Y values in the training data, conditioned on their corresponding X values.

CausFunc₀: the Best Algorithm is JRip

For **CausFunc₀**, JRip has shown precision of 0.747, recall of 0.705, and F-measure of 0.725.

JRip (Extended Repeated Incremental Pruning) implements a propositional rule learner, "Repeated Incremental Pruning to Produce Error Reduction" (RIPPER), as proposed in (Cohen, 1995). JRip is a rule learner alike in principle to the commercial rule learner RIPPER.

CausFunc₁: the Best Algorithm is J48

For **CausFunc₁**, J48 has shown precision of 0.842, recall of 0.696, and F-measure of 0.762.

J48 is a rule base classifier algorithm that generates C4.5 decision trees. J48 is the C4.5 clone implemented in the WEKA data mining library. In its turn, C4.5 implements ID3 algorithm.

The basic ideas behind ID3 are the following. Firstly, in the decision tree each node corresponds to a non-categorical attribute and each arc to a possible value of that attribute. A leaf of the tree specifies the expected value of the categorical attribute for the records described by the path from the root to that leaf. Secondly, in the decision tree at each node should be associated the non-categorical attribute which is most informative among the attributes not yet considered in the path from the root. Thirdly, entropy is used to measure how informative is a node. C4.5 is an extension of ID3 that accounts for unavailable values, continuous attribute value ranges, pruning of decision trees, rule derivation, etc.

Real₁: the Best Algorithm is Prism

For **Real₁**, Prism has shown precision of 0.735, recall of 0.832, and F-measure of 0.781.

Prism is a rule based classification algorithms It is based on the inductive rule learning and uses separate-and-conquer strategy. It means, that a rule that works for many instances in the class is identified first, then the instances covered by this rule are excluded from the training set and the learning continues on the rest of the instances. These learners are efficient on large, noisy datasets. Our training sets included 900 instances represented as vectors of the size 1109 attributes, and rule induction algorithms performed very well.

Func$_0$: the Best Algorithm is BFTree

For **Func$_0$**, BFTree has shown precision of 0.667, recall of 0.727, and F-measure of 0.696.

BFTree is a best-first decision tree learner and it is a learning algorithm for supervised classification learning. Best-first decision trees represent an alternative approach to standard decision tree techniques such as C4.5 algorithm since they expand nodes in best-first order instead of a fixed depth-first order.

Oper$_2$: the Best Algorithm is PART

For **Oper$_2$**, PART has shown precision of 0.923, recall of 0.571, and F-measure of 0.706.

PART is a rule base classifier that generates partial decision trees rather that forming one whole decision tree in order to achieve the classification.

IncepOper$_1$: the Best Algorithm is Prism

For **IncepOper$_1$**, Prism has shown precision of 0.750, recall of 0.800, and F-measure of 0.774.

Since Prism is also the best classifier for **Real$_1$**, it has been described above. We remind, that for **Real$_1$**, Prism has shown precision of 0.735, recall of 0.832, and F-measure of 0.781. **IncepOper$_1$** has the meaning 'to begin doing what is designated by the noun', and **Real$_1$** means 'to act according to the situation designated by the noun'. The meaning of **IncepOper$_1$** is distinguished better because the semantic element 'begin' is very different from 'act' which is more general and has more resemblance with the meaning 'perform' (**Oper$_1$**) or 'cause to exist' (**CausFunc$_0$**, **CausFunc$_1$**).

ContOper$_1$: the Best Algorithm is SimpleLogistic

For **ContOper$_1$**, SimpleLogistic has shown precision of 0.909, recall of 0.769, and F-measure of 0.833.

Simple logistic regression finds the equation that best predicts the value of the Y variable (in our case, the class attribute) for each value of the X variable (in our training data, values of attributes that represent hypernyms). What makes logistic regression different from linear regression is that the Y variable is not directly measured; it is instead the probability of obtaining a particular value of a nominal variable.

The algorithm calculates the probability of Y applying the likelihood ratio method. This method uses the difference between the probability of obtaining the observed results under the logistic model and the probability of obtaining the observed results in a model with no relationship between the independent and dependent variables.

FWC: the Best Algorithm is Prism

For **FWC**, Prism has shown precision of 0.639, recall of 0.702, and F-measure of 0.669.

Since Prism is also the best classifier for **Real$_1$**, it has been described above. Prism has the top result for **IncepOper$_1$** as well. Still, for free word combinations, Prism show weaker results that for predicting **IncepOper$_1$** and **Real$_1$**. We remind that for **IncepOper$_1$**, Prism has shown precision of 0.750, recall of 0.800, and F-measure of 0.774; and for **Real$_1$**, this algorithm has shown precision of 0.735, recall of 0.832, and F-measure of 0.781.

The meaning of free word combinations is less distinguishable since their semantic content is very diverse in comparison with the meaning of lexical functions.

5.5.6 LF Identification k-Class Classification Problem

The task of annotating Spanish verb-noun collocations with lexical funcitons can also be viewed as a k-class classification problem. For this experiment, we chose seven lexical funcitons: **Oper$_1$**, **CausFunc$_1$**, **IncepOper$_1$**, **ContOper$_1$**, **Func$_0$**, **Real$_1$**, and **Oper$_2$**.

Each lexical funcitons is seen as a category, thus we had 7-class classification. To perform such classification, we chose a number of methods which may be called characteristic of various commonly used approaches in machine learning: Bayesian classification, rule induction, decision tree construction, the nearest neighbor technique, and kernel methods. For experimentation, the following classifiers implemented in WEKA have been chosen: **NaiveBayes** as a classical probabilistic Bayesian classifier; **PART**, **JRip**, **Prism**, **Ridor** for rules; **BFTree**, **SimpleCart**, **FT**, **REPTree** for trees; **IB1** a basic nearest neighbor instance based learner using one nearest neighbor for classification, and **SMO** (Sequential Minimal Optimization), an implementation of support vector machine, for kernel techniques.

Table 15 present the results of these experiments which are values of F-measure calculated according to the formula in Section 5.2 Also, for algorithms representing rule induction (PART, JRip, Prism, and Ridor), the number of rules created in a respective classification model is given in the first column of Table 15.

The best result in Table 15 is shown by SMO. This algorithm was able to reach the F-measure of 0.916 for predicting **Oper$_1$**. It is also the best among methods indicated in Table 15 for the meanings **IncepOper$_1$** and **Func$_0$**. SMO achieved the second best weighted average for all seven generalized meanings.

Table 15 Performance of some algorithms using 7-class approach

Algorithm	$Oper_1$	$CausFunc_1$	$IncepOper_1$	$ContOper_1$	$Func_0$	$Real_1$	$Oper_2$	Weighted average
rules.PART: 21 rules	0.894	0.783	0.524	0.774	0.903	0.685	0.643	0.812
rules.JRip: 26 rules	0.878	0.800	0.634	0.800	0.903	0.686	0.667	0.815
rules.Prism: 222 rules	0.896	0.840	0.647	0.720	0.889	0.744	0.682	0.841
rules.Ridor: 31 rules	0.888	0.709	0.667	0.774	0.710	0.576	0.618	0.780
trees.BFTree	0.908	0.814	0.605	0.800	0.875	0.672	0.667	0.830
trees.SimpleCart	0.915	0.798	0.605	0.800	0.875	0.672	0.656	0.829
trees.FT	0.915	**0.863**	0.714	**0.875**	0.903	**0.757**	**0.733**	**0.865**
trees.REPTree	0.893	0.746	0.632	0.759	0.759	0.529	0.677	0.788
bayes.NB	0.759	0.698	0.000	0.000	0.000	0.119	0.000	0.549
lazy.IB1	0.783	0.620	0.378	0.519	0.688	0.462	0.444	0.664
functions.SMO	**0.916**	0.843	**0.773**	0.839	**0.933**	0.739	0.714	0.861

As it is previously mentioned, SMO is an implementation of support vector machine (Cortes and Vapnik, 1995), a non-probabilistic binary linear classifier. For a given instance of training data, it predicts which of two possible classes the instance belongs to. A support vector machine model is a representation of the examples as points in space, mapped so that the examples of the separate classes are divided by a clear gap. New examples are then mapped into that same space and predicted to belong to a category based on which side of the gap they fall on.

SVM have been used successfully for many NLP tasks, for example, word sense disambiguation, part of speech tagging, language identification of names, text categorization, and others. As our results demonstrate, it is also effective for annotating collocations with lexical functions.

However, FT (Functional Tree), a generalization of multivariate trees able to explore multiple representation languages by using decision tests based on a combination of attributes (Gama, 2004), was more successful than SMO for predicting the meanings $CausFunc_1$, $ContOper_1$, $Real_1$, and $Oper_2$, for which FT acquired the highest value of F-measure.

For rule induction algorithms, Table 4 in Appendix 1 includes the number of rules generated by each technique. It appears that PART and JRip are more effective since they show high values of F-measure, 0.812 and 0.815, respectively, and generate quite a modest number of rules (21 and 26) compared with Prism which in spite of a higher F-measure of 0.841 generates as much as 222 rules.

NaiveBayes (Jiang *et al.*, 2010) and IB1 (Aha and Kibler, 1991) showed a well-noted tendency to have low F-measure for all meanings in the experiments. The failure of NaiveBayes may be explained by the fact that statistical methods are limited by the assumption that all features in data are equally important in contributing to the decision of assigning a particular class to an example and also independent of one another.

However, it is a rather simplistic view of data, because in many cases data features are not equally important or independent. The latter is certainly true for linguistic data, especially for such a language phenomenon as hypernyms (recall that the meaning of collocations is represented by hypernyms in our training sets).

Hypernyms in the Spanish WordNet form a hierarchic structure where every hypernym has its ancestor, except for the most general hypernyms at the top of the hierarchy, and daughter(s), except for most specific hypernyms at the end of the hierarchy.

In spite of these problems, Naive Bayes is one of the most common algorithms used in natural language processing, it is effective in text classification (Eyheramendy *et al.*, 2003), word sense disambiguation (Pedersen, 2000), information retrieval (Lewis, 1998). In spite of that, it could hardly distinguish the generalized meanings of collocations in our experiments. In the previous paragraph some reasons for this failure are suggested.

Low results of IB1 demonstrates that normalized Euclidean distance used in this technique to find the training instance closest to the given test instance does not approximate well the target classification function. Another reason of low performance can be the fact that if more than one instances have the same smallest distance to the test instance under examination, the first one found is used, which can be erroneous.

Since all best classifiers for predicting the generalized meaning are rule-based, we can suppose that semantics of collocations is better distinguished by rules than on the basis of probabilistic knowledge learned from the training data.

Chapter 6
Linguistic Interpretation

Linguistics as a scientific study of human language intends to describe and explain it. However, validity of a linguistic theory is difficult to prove due to volatile nature of language as a human convention and impossibility to cover all real-life linguistic data. In spite of these problems, computational techniques and modeling can provide evidence to verify or falsify linguistic theories.

As a case study, we conducted a series of computer experiments on verb-noun collocations using machine learning methods described in Chapter 5, in order to test a linguistic point that collocations in the language do not form an unstructured collection but are language items related via what we call collocational isomorphism, represented by lexical functions of the Meaning-Text Theory. Our experiments allowed us to verify this linguistic statement. Moreover, they suggested that semantic considerations are more important in the definition of the notion of collocation than statistical ones.

6.1 Computer Experiments in Linguistics

Computer experiments play a very important role in science today. Simulations on computers have not only increased the demand for accuracy of scientific models, but have helped the researcher to study regions which can not be accessed in experiments or would demand very costly experiments.

Our computational experiments made on the material of Spanish verb-noun collocations like *seguir el ejemplo*, follow the example, *satisfacer la demanda*, meet the demand, *tomar una decisión*, make a decision, can contribute to verify if the linguistic statement specified in Section 6.2 is true. Out results also make it possible to derive another important inference on the nature of collocation presented in Section 6.4.

It should be added here that testing a linguistic hypothesis on computer models not only demonstrates validity or rejection of the hypothesis, but also motivates the researcher to search for more profound explanations or to explore new approaches in order to improve computational operation. Thus, starting from one linguistic model, the researcher can evaluate it and then go further, sometimes into neighboring spheres of linguistic reality, in her quest of new solutions, arriving at interesting conclusions. One of the original intents of our research was to test one linguistic model experimentally.

A. Gelbukh, O. Kolesnikova: Semantic Analysis of Verbal Collocations, SCI 414, pp. 85–92.

If we develop a computer program on the premise of a certain linguistic model and this program accomplishes its task successfully, then the linguistic model being the basis of the program is thus verified. The results we obtained not only produced evidence for verifying a linguistic model or statement we make in the next section, but they also made it possible to get more insight into the nature of collocation which has been a controversial issue in linguistics for many years.

6.2 Linguistic Statement: Our Hypothesis

Collocations are not a stock or a "bag" of word combinations, where each combination exists as a separate unit with no connection to the others, but they are related via collocational isomorphism represented as lexical functions.

6.2.1 Collocational Isomorphism

Considering collocations of a given natural language, it can be observed that collocations are not just a "bag" of word combinations, as a collection of unrelated items where no association could be found, but there are lexical relations among collocations, and in particular, we study the lexical relation which may be called 'collocational isomorphism'. It has some resemblance to synonymy among words which is the relation of semantic identity or similarity. Collocational isomorphism is not a complete equality of the meaning of two or more collocations, but rather a semantic and structural similarity between collocations.

What do we mean by semantic and structural similarity between collocations? For convenience of explanation, we will comment on the structural similarity of collocations first. The latter is not a novelty, and a detailed structural classification of collocations (for English) was elaborated and used to store collocational material in the well-known dictionary of word combinations *The BBI Combinatory Dictionary of English* (Benson *et al.* 1997). However, we will exemplify collocational structures with Spanish data, listing some typical collocates of the noun *alegría*, joy:

> verb + noun: *sentir alegría*, to feel joy
> adjective + noun: *gran alegría*, great joy
> preposition + noun: *con alegría*, with joy
> noun + preposition: *la alegría de* (esa muchacha), the joy of (this girl).

The above examples are borrowed from the dictionary of Spanish collocations entitled *Diccionario de colocaciones del español* (Alonso *et al.*, 2010; Vincze *et al.*, 2011), a collection of collocations in which the bases are nouns belonging to the semantic field of emotions. So collocations have structural similarity when they share a common syntactic structure.

We say that two or more collocations are similar semantically if they possess a common semantic content. In Table 16, we present collocations with the same syntactic structure, namely, "verb + noun". For these collocations, the meaning is given for us to see what semantic element can be found that is common to all of them.

Table 16 Verb-noun collocations and their meaning

Collocations	Meaning of collocation
make use	*to use*
give a kiss	*to kiss*
have a look	*to look*
feel joy	*to rejoice*
experience grief	*to grieve*

Table 17 Verb-noun collocations grouped according to their common semantic pattern

Semantic pattern	Collocations
create an entity or process	*write a book* *develop a plan* *build a society* *give life*
intensify a property or attribute	*increase the risk* *raise the level* *develop a capacity* *improve a condition*
reduce a property or attribute	*lower chances* *reduce consumption* *bring down the price* *restrict rights*
begin to realize an action or to manifest an attribute	*open a session* *adopt an attitude* *take the floor* *assume a role*
preserve a property or process	*maintain the balance* *keep silent* *follow an example* *lead a life*

It may be noted that the meaning of all collocations in Table 16 is generalized as 'do, carry out or realize what is denoted by the noun', in other words, that these collocations are built according to the semantic pattern 'do the noun'. In turn, observing the meaning of the nouns, we see that their semantics can be expressed in general terms as 'action' (*use, kiss, look*) or 'psychological feature' (*joy, grief*), so the resulting semantic pattern of the collocations in Table 16 is 'do an action / manifest a psychological feature'. Since these collocations share common semantics and structure, we may say that they are isomorphic, or that they are tied to one another by the relation we termed above as 'collocational isomorphism'. Table 17 gives more examples of isomorphic collocations.

6.2.2 Collocational Isomorphism Represented as Lexical Functions

Several attempts to conceptualize and formalize semantic similarity of collocations have been made. As far back as in 1934, the German linguist W. Porzig claimed that on the syntagmatic level, the choice of words is governed not only by grammatical rules, but by lexical compatibility, and observed semantic similarity between such word pairs as *dog – bark*, *hand – grasp*, *food – eat*, *cloths – wear* (Porzig, 1934). The common semantic content in these pairs is 'typical action of an object'.

Research of J. R. Firth (1957) drew linguists' attention to the issue of collocation and since then collocational relation has been studied systematically. In the article of J. H. Flavell and E. R. Flavell (1959) and in the paper by Weinreich (1969), there were identified the following meanings underlying collocational isomorphism: an object and its typical attribute (*lemon – sour*), an action and its performer (*dog – bark*), an action and its object (*floor – clean*), an action and its instrument (*axe – chop*), an action and its location (*sit – chair*, *lie – bed*), an action and its causation (*have – give*, *see – show*), etc. Examples from the above mentioned writings of Porzig, Flavell and Flavell, Weinreich are borrowed from (Apresjan, 1995: 44).

The next step in developing a formalism representing semantic relations between the base and the collocate as well as semantic and structural similarity between collocations was done by I. Mel'čuk. Up to now, his endeavor has remained the most fundamental and theoretically well-grounded attempt to systematize collocational knowledge. This scholar proposed a linguistic theory called the Meaning-Text Theory, which explained how meaning, or semantic representation, is encoded and transformed into spoken or written texts (Mel'čuk, 1974). His theory postulates that collocations are produced by a mechanism called lexical function. Lexical function is a mapping from the base to the collocate; it is a semantically marked correspondence that governs the choice of the collocate for a particular base. About 70 lexical functions have been identified in (Mel'čuk, 1996); each is associated with a particular meaning according to which it receives its name. Table 18 demonstrates a correspondence between semantic patterns of collocations and lexical functions.

Now we are going to see if computer experiments can supply evidence to the existence of collocational isomorphism as defined by lexical functions. The idea is to submit a list of collocations to the computer and see if it is able to distinguish collocations belonging to different lexical functions. If a machine can recognize lexical functions, then it is a strong evidence of their existence.

Table 18 Semantic patterns represented as lexical functions

Semantic pattern and examples	Complex lexical function representation	Complex lexical function description
create an entity or process *write a book* *give life*	$\text{CausFunc}_0(book) = write$ $\text{CausFunc}_0(life) = give$	$\text{CausFunc}_0 = $ cause an entity or process to function.
intensify a property or feature *increase the risk* *raise the level*	$\text{CausPlusFunc}_1(risk) = increase$ $\text{CausPlusFunc}_1(level) = raise$	$\text{CausPlusFunc}_1 = $ cause that a property or feature manifest itself to a larger degree.
reduce a property or feature *lower chances* *reduce consumption*	$\text{CausMinusFunc}_1(chances) = lower$ $\text{CausMinusFunc}_1(consumption) = reduce$	$\text{CausMinusFunc}_1 = $ cause that a property or feature manifest itself to a lesser degree.
begin to realize an action or begin to manifest a feature *open a session* *adopt an attitute*	$\text{IncepOper}_1(session) = open$ $\text{IncepOper}_1(attitude) = adopt$	$\text{IncepOper}_1 = $ begin to realize an action or to manifest a feature.
preserve a property or process *maintain the balance* *keep silent*	$\text{ContOper}_1(balance) = maintain$ $\text{ContOper}_1(silent) = keep$	$\text{ContOper}_1 = $ continue to realize an action or to manifest a feature.

6.3 Testing the Linguistic Statement

One of the purposes of this work is to provide evidence for the linguistic statement made in the beginning of Section 6.2. Now let us review it in the light of our experimental results. The statement affirms that collocations are not a stock, or a "bag" of word combinations, where each combination exists as a separate unit with no connection to others, but they are related via collocational isomorphism represented as lexical functions.

What evidence have we obtained concerning lexical functions? We presented a sufficient number of collocations annotated with lexical functions to the computer that learned characteristic features of each function. It was demonstrated that the computer was able to assign lexical functions to unseen collocations with a significant average accuracy of 0.758. Is it satisfactory?

We can compare our result with computer performance on another task of natural language processing: word sense disambiguation, i.e., identifying the intended meanings of words in context. Today, automated disambiguating systems reach the accuracy of about 0.700 and this is considered a substantial achievement. As an example of such works see (Zhong and Tou Ng, 2010). Therefore, our result is weighty enough to be a trustworthy evidence for the linguistic statement under discussion.

In Section 6.1 we stated, that if we develop a computer program on the premise of a certain linguistic model and this program accomplishes its task successfully, then the linguistic model being the basis of the program is thus verified. In our experiments, we have observed that machine learning methods are able to detect lexical functions of collocations. Thus lexical functions as a linguistic concept get evidence received in computational experiments which can be repeated on the same data as well as on new data. It means that the formalism of lexical functions is a legitimate model of collocational isomorphism described in Section 6.2.

6.4 Statistical and Semantic View on Collocation

What knowledge is necessary and sufficient for the computer to analyze and generate texts in natural language? And what type of knowledge should it be? Up to now, the two foremost approaches in natural language processing have been the statistical approach and the symbolic one. Our results demonstrated that rule-based methods outperformed statistical methods in detecting lexical functions. It means that collocations are analyzed better by rules than by frequency counts; that rules tell us more of what collocations are than frequency counts do; that collocations can be recognized better semantically than statistically.

The fact that the semantic aspect of collocation outweighs the statistical one has an important effect on the definition of collocations. Definition of a concept must contain necessary and sufficient criteria for distinguishing this concept from other concepts.

The debate over the most relevant criterion for defining collocations has already lasted over a long period. Should this criterion be statistical or semantic? Wanner (2004) gives a good concise overview of this debate. The statistical definition of collocation, i.e. based on probabilistic knowledge, says that collocation is the syntagmatic association of lexical items, quantifiable, textually, as the probability that there will occur, at n removes (a distance of n lexical items) from an item x, the items a, b, c ... (Halliday, 1961:276).

The semantic definition of collocation explains how the collocational meaning is formed: a collocation is a combination of two words in which the semantics of the base is autonomous from the combination it appears in, and where the collocate adds semantic features to the semantics of the base (Mel'čuk, 1995). For example, in the phrase *She fell to the floor*, all the words are used in their typical sense and the verb *to fall* means *to drop oneself to a lower position*, but when it is said *She fell in love*, we understand that the same verb is not used in its typical, full meaning, but attains a different sense 'begin to experience something'. *WordReference* Online Dictionary (www.wordreference.com) gives a description of this sense: pass suddenly and passively into a state of body or mind.

To illustrate the definition, the dictionary referenced in the previous paragraph provides the following examples: *to fall into a trap, She fell ill, They fell out of favor, to fall in love, to fall asleep, to fall prey to an imposter, fall into a strange way of thinking*. This meaning of *fall* is more abstract as compared with its typical meaning given in WordNet 'descend in free fall under the influence of gravity', e.g., *The branch fell from the tree*. *Fall* reveals its characteristic meaning in free word combinations, and its more abstract sense, in collocations.

What do we mean by more abstract sense? An abstract sense is not independent, it is not complete, but rather can be called a "semantic particle" whose function is not to express the full semantics, but to add semantic features to the base of collocation.

To explain what is meant by "adding semantic features to the base", let us make an analogy with semantics of grammatical categories which is also very abstract. The verb *be* in its function as an auxiliary verb does not express any meaning except abstract grammatical categories of time, aspect, and person. In the sentence *This castle was built in the 15th century*, the verb *build* carries the meaning of an action, and what *be* does is adding semantic features to the verb, i.e. that this action took place in past, it is passive, not active, and was applied to a single object, because the grammatical number of *be* is singular. Likewise, *fall* does not express an event, or a state, but "adds" to the word denoting an event or state the semantic feature 'begin to occur'.

According to the semantic definition of collocation, the latter differs from free word combinations in the way it constructs its semantics. While the semantics of a free word combination is the sum of the meanings of its elements, collocational meaning is formed by adding more abstract semantic features expressed by the collocate to the full meaning of the base.

Our experiments showed that collocations are recognized better using rules, or conceptual knowledge. It means that the basic criterion for distinguishing collocations from free word combinations is semantic, so there is a good evidence and reason to build definition of collocation on the semantic, not statistical, criterion.

6.5 Conclusions

The following conclusions can be drawn from the above analysis:

- Our experiments have shown that verb-noun collocations can be classified according to semantic taxonomy of lexical functions using supervised machine learning tecniques. We have shown that it is feasible to apply machine learning methods for predicting the meaning of unseen Spanish verb-noun collocations.
- Verb-noun pairs were represented as sets of hypernyms for both the verb and the noun. As our experiments have shown, hypernyms of the Spanish WordNet function sufficiently well as features distinguishing between the meanings we chosen to be predicted by classifiers. Therefore, this representation can be used for the task of automatic extraction of lexical functions. With this we re-confirmed that the set of hypernyms can be used to describe lexical meaning and discriminate word senses.
- According to 10-fold cross-validation technique, the best performance was demonstrated by bayes.BayesianLogisticRegression algorithm for detecting

the lexical function **Oper**$_1$ and by SimpleLogistic classifier for detecting the lexical function **ContOper**$_1$. Both algorithms can be applied for high quality semantic annotation of verb-noun collocations based on the taxonomy of lexical functions.

– According to evaluation of algorithms on an independent test set, the best performance was shown by Ridor and LWL algorithms for detecting the lexical function **ContOper**$_1$. These algorithms can be used for high quality annotation of verb0noun collocations with this lexical function.

– The best f-measure achieved in our experiments is 0.873 using the training set and 10-fold cross-validation technique. This is significantly higher than the previously reported result of 0.740 for F-measure, though the comparison is not fair because we looked for the meaning which is similar to the meaning predicted in (Wanner *et al.*, 2006), but not the same one. The highest F-measure achieved in the experiments on an independent test set was only 0.658. This could be explained by the fact that the best ratio between the training set and the test set has not yet been found by us. More experiments on test sets of various sizes are needed.

– We have tested if the three hypothesis stated in (Wanner *et al.*, 2006) were valid and well-grounded. These hypothesis claim that collocations can be recognized: first, by their similarity to the prototypical sample of each collocational type (this strategy is modeled by the Nearest Neighbor technique); second, by similarity of semantic features of their elements (i.e., base and collocate) to semantic features of elements of the collocations known to belong to a specific LF (this method is modeled by Naïve Bayesian network and a decision tree classification technique based on the ID3-algorithm); and third, by correlation between semantic features of collocational elements (this approach is modeled by Tree-Augmented Network Classification technique). Our research has shown that there machine learning methods other than mentioned in the three hypotheses that can be used for high quality annotation of verb-noun collocations of lexical funciton. To these methods the following algorithms belong: JRip, J48, Prism, PART, SimpleLogistic, Ridor.

Chapter 7
Dictionary of Spanish Verbal Lexical Functions

In this chapter, we describe a lexical resource compiled by us and used in the experiments presented in Chapter 5.

7.1 Compilation

Firstly, the Spanish Web Corpus available via the SketchEngine, online software for natural language processing (Kilgarriff *et al.*, 2004), was chosen as a source of verb-noun pairs with the patterns "verb + complement" and "noun + verb" (in Spanish sentences, the verb functioning as the predicate often preceeds the noun as the subject, therefore, such combinations were extracted as verb-noun pairs; in our dictionary they are transaformed into a standard form "noun + verb"). All such verb-noun pairs used in the Spanish Web Corpus five or more times, were extracted automatically from the said corpus.

Fig. 11 displays the interface of the Sketch Engine where several corpora are listed including the Spanish Web Corpus. The obtained list contained 83,982 verb-noun pairs, and it was ranked by frequency. The software in on the SketchEngine website: www.sketchengine.co.uk.

Secondly, one thousand pairs were taken from the upper part of the list, i.e. most frequent verb-noun pairs.

Thirdly, in the list of one thousand pairs, erroneous combinations were marked with the label ERROR. Erroneous pairs included, for instance, past participle or infinitive instead of noun, or contained symbols like --, « , © instead of words. How did errors emerge? The automatic extraction procedure was set to search for combinations with the pattern "verb + complement" in the corpus. This procedure needs part of speech (POS) and lemma information, and such data is supplied by TreeTagger, software used to annotate the Spanish Web Corpus with POS and lemmas.

The TreeTagger is a leading tool applied for POS tagging and lemmatisation, it achieves high accuracy but still is error-prone. Due to errors made by the TreeTagger, the set of extracted verb-noun pairs contained fallacious combinations. For the sake of preserving the original design of automatically extracted set, these incorrect combinations were not removed from the list but identified as wrong. The total number of erroneous pairs was 61, so after their removal the list contained 939 pairs.

A. Gelbukh, O. Kolesnikova: Semantic Analysis of Verbal Collocations, SCI 414, pp. 93–96.
springerlink.com © Springer-Verlag Berlin Heidelberg 2013

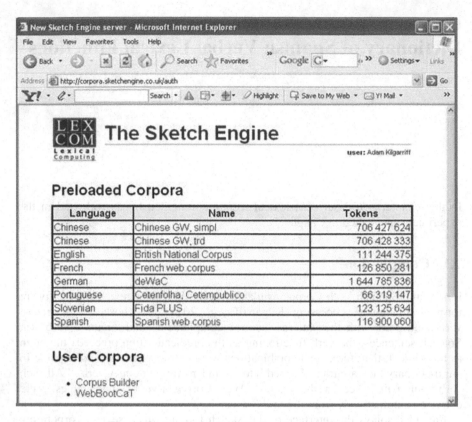

Fig. 11 Sketch Engine with the Spanish Web Corpus.

Fourthly, collocational verb-noun pairs were annotated with lexical functions. The rest of the pairs were annotated as free word combinations using the label FWC.

Lastly, all verbs and nouns in the list were disambiguated with word senses from the Spanish WordNet, an electronic lexicon structured the same way as WordNet for English. For some verb-noun pairs, relevant senses were not found in the above mentioned dictionary, and the number of such pairs was 39.

For example, in the combinaiton *dar cuenta* (to give account), the noun *cuenta* means *razón, satisfacción de* algo (reason, satisfaction of something). This sense of *cuenta* is taken from Diccionario de la Lengua Española (Dictionary of the Spanish Language, 2001). Unfortunately, this sense is absent in the Spanish WordNet so the expression *dar cuenta* was left without sense annotation. All such words were annotation N/A, i.e. not available.

The annotated list was formatted as a table and saved in an MS Excel file. Fig.12 shows the process of the compilation of the lexical resource schematically.

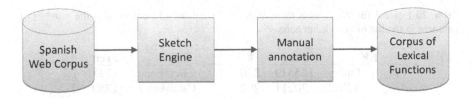

Fig. 12 The process of lexical resource compilation.

7.2 Description

A partial representation of the list is given in Table 19; Table 20 lists all lexical functions found in the list of 1000 most frequent verb-noun pairs, their frequencies in the Spanish Web Corpus, and the number of examples for each of them.

Table 19 Partial representation of the lexical resource

LF/ FWC/ ERROR	Verb	Verb Sense Number	Noun	Noun Sense Number	Frequency
$Oper_1$	*dar*	2	*cuenta*	N/A	9236
$CausFunc_0$	*formar*	2	*parte*	1	7454
$Oper_1$	*tener*	1	*lugar*	4	6680
$Oper_1$	*tener*	1	*derecho*	1	5255
$CausFunc_1$	*hacer*	2	*falta*	N/A	4827
$CausFunc_1$	*dar*	9	*lugar*	4	4180
$Oper_1$	*hacer*	15	*referencia*	2	3252
$Func_0$	*hacer*	N/A	*año*	2	3211
$Oper_1$	*tener*	1	*problema*	7	3075
$Func_0$	*hacer*	N/A	*tiempo*	1	3059
$IncepOper_1$	*tomar*	4	*decisión*	2	2781
$Oper_1$	*tener*	1	*acceso*	3	2773
$Oper_1$	*tener*	1	*razón*	2	2768
$Caus_2Func_1$	*llamar*	8	*atención*	1	2698
$Oper_1$	*tener*	1	*sentido*	1	2563
ERROR	*haber*		*estado*		2430
FWC	*hacer*	6	*cosa*	3	2374
$Oper_1$	*tener*	3	*miedo*	1	2226
ERROR	*haber*		*hecho*		2168

Table 20 Lexical functions with their respective frequency in corpus and the number of instances in the list of verb-noun pairs

LF	Freq	#	LF	Freq	#
$Oper_1$	165319	280	$PerfFunc_0$	1293	1
FWC	70211	202	$Caus_1Oper_1$	1280	2
$CausFunc_1$	45688	90	$Caus_1Func_1$	1085	3
$CausFunc_0$	40717	112	$IncepFunc_0$	1052	3
ERROR	26316	61	$PermOper_1$	910	3
$Real_1$	19191	61	$CausManifFunc_0$	788	2
$Func_0$	17393	25	$CausMinusFunc_0$	746	3
$IncepOper_1$	11805	25	$Oper_3$	520	1
$Oper_2$	8967	30	$LiquFunc_0$	514	2
$Caus_2Func_1$	8242	16	$IncepReal_1$	437	2
$ContOper_1$	5354	16	$Real_3$	381	1
Manif	3339	13	$PlusOper_1$	370	1
Copul	2345	9	$CausPerfFunc_0$	290	1
$CausPlusFunc_0$	2203	7	$AntiReal_3$	284	1
$Func_1$	1848	4	$MinusReal_1$	265	1
$PerfOper_1$	1736	4	$AntiPermOper_1$	258	1
$CausPlusFunc_1$	1548	5	$ManifFunc_0$	240	1
$Real_2$	1547	3	$CausMinusFunc_1$	229	1
$FinOper_1$	1476	6	$FinFunc_0$	178	1

The complete dictionary is given in Appendix 2.

References

Aha, D., Kibler, D.: Instance-based learning algorithms. Machine Learning 6, 37–66 (1991)

Alonso Ramos, M., Rambow, O., Wanner, L.: Using semantically annotated corpora to build collocation resources. In: Proceedings of LREC, Marrakesh, Morocco, pp. 1154–1158 (2008)

Alonso, M., Nishikawa, A., Vincze, O.: DiCE in the web: An online Spanish collocation dictionary. In: Granger, S., Paquot, M. (eds.) eLexicograpy in the 21st Century: New Challenges, New Applications. Proceedings of eLex 2009, Cahiers du Cental 7, pp. 367–368. Presses universitaires de Louvain, Louvain-la-Neuve (2010)

Apresjan, J.D.: Lexical Semantics. Vostochnaya Literatura RAN, Moscow (1995) (in Russian)

Apresjan, J.D.: Systematic Lexicography. Oxford University Press, US (2008); Translated by Windle, K.

Apresjan, J.D., Boguslavsky, I.M., Iomdin, L.L., Tsinman, L.L.: Lexical Functions as a Tool of ETAP-3. In: MTT 2003, Paris, June 16-18 (2003)

Apresian, J., Boguslavsky, I., Iomdin, L., Lazursky, A., Sannikov, V., Sizov, V., Tsinman, L.: ETAP-3 Linguistic Processor: a Full-Fledged NLP Implementation of the MTT. In: First International Conference on Meaning-Text Theory, MTT 2003, pp. 279–288. Ecole Normale Superieure, Paris (2003)

Benson, M.: Collocations and general-purpose dictionaries. International Journal of Lexicography 3(1), 23–35 (1990)

Benson, M., Benson, E., Ilson, R.: The BBI Combinatory Dictionary of English. John Benjamins, Amsterdam (1986)

Brinton, L.J., Brinton, D.M.: The Linguistic Structure of Modern English. John Benjamins Publishing Company, Amsterdam (2010)

Burchardt, A., Erk, K., Frank, A., Kowalski, A., Padó, S., Pinkal, M.: The SALSA corpus: a German corpus resource for lexical semantics. In: Proceedings of LREC, Genova, Italy (2006)

Chiu A., Poupart P. and DiMarco C.: Learning Lexical Semantic Relations using Lexical Analogies, Extended Abstract. Publications of the Institute of Cognitive Science (2007)

Cohen, W.: Fast effective rule induction. In: 12th International Conference on Machine Learning, pp. 115–123 (1995)

Coppin, B.: Artificial Intelligence Illuminated. Jones and Bartlett Publishers, Sudbury (2004)

Cortes, C., Vapnic, V.: Support Vector Networks. Machine Learning 20, 1–25 (1995)

Cowie, A.P.: Phraseology. In: Asher, R.E. (ed.) The Encyclopedia of Language and Linguistics. Pergamon Press, Oxford (1994)

Diachenko, P.: Lexical functions in learning the lexicon. Current Developments in Technology-Assisted Education 1, 538–542 (2006)

Diccionario de la lengua española. Real Academia Española, 22nd edn. Real Academia
    Española, Madrid (2001)
Firth, J.R.: Modes of Meaning. In: Firth, J.R. (ed.) Papers in Linguistics 1934–1951, pp.
    190–215. Oxford University Press, Oxford (1957)
Flavell, J.H., Flavell, E.R.: One Determinant of Judged semantic and associative connection
    between words. Journal of Experimental Psychology 58(2), 159–165 (1959)
Gama, J.: Functional Trees. Machine Learning 55(3), 219–250 (2004)
Gledhill, C.J.: Collocations in Science Writing. Gunten Narr Verlag, Tübingen (2000)
Fontenelle, T.: What on Earth are Collocations? English Today 10(4), 42–48 (1994)
Hall, M., Frank, E., Holmes, G., Pfahringer, B., Reutemann, P., Witten, I.H.: The WEKA
    Data Mining Software: An Update. SIGKDD Explorations 11(1) (2009)
Halliday, M.A.K.: Categories of the Theory of Grammar. Word 17, 241–292 (1961)
Hausmann, F.J.: Wortschatzlernen ist Kollokationslernen. Zum Lehren und Lernen französischer
    Wortverbindungen. Praxis des neusprachlichen Unterrichts 31, 395–406 (1984)
Howarth, P.: Phraseology in English academic writing. Some Implications for Language
    Learning and Dictionary Making. Tübingen, Niemeyer (1996)
Iomdin, L., Cinman, L.: Lexical Functions and Machine Translation. In: Dialogue 1997:
    Proceedings of the Dialogue International Conference on Computational Linguistics
    and its Applications (1997)
Jiang, L., Cai, Z., Wang, D.: Improving naive Bayes for classification. International Journal
    of Computers and Applications 32(3), 328–332 (2010)
Kilgarriff, A., Rychly, P., Smrz, P., Tugwell, D.: The Sketch Engine. In: Proceedings of
    EURALEX, pp. 105–116. Université de Bretagne Sud, France (2004)
Leed, R.L., Nakhimovsky, A.D.: Lexical Functions and Language Learning. Slavic and
    East European Journal 23(1), 104–113 (1979)
Leed, R.L., Iordanskaja, L., Paperno, S., et al.: A Russian-English Collocational Dictionary
    of the Human Body (1996)
Lewis, M.: The Lexical Approach. The State of ELT And A Way Forward. Language
    Teaching Publications (1994)
Lewis, M.: Implementing the Lexical Approach: Putting Theory into Practice. Language
    Teaching Publications, Hove (1997)
Little, W., Senior Coulson, J., Onions, C.T., Fowler, H.W.: The Shorter Oxford English
    Dictionary on Historic Pronciples. Clarendon Press, Oxford (1959)
Loos, E., Anderson, S., Day Jr., D.H., Jordan, P.C., Wingate, J.D. (eds.): Glossary of
    Linguistic Terms (1997), http://www.sil.org/linguistics/
    GlossaryOfLinguisticTerms/ (last viewed on June 08, 2010)
Malmkjær, K. (ed.): The Linguistics Encyclopedia, 2nd edn. Routledge, London (2002)
McIntosh, C., Francis, B., Poole, R. (eds.): Oxford Collocations Dictionary for Students of
    English. Oxford University Press, Oxford (2009)
Mel'čuk, I.: Opyt teorii lingvističeskix modelej "Smysl ↔ Tekst" ['A Theory of the
    Meaning-Text Type Linguistic Models']. Nauka, Moskva (1974)
Mel'čuk, I.: Lexical Functions: A Tool for the Description of Lexical Relations in a
    Lexicon. In: Wanner, L. (ed.) Lexical Functions in Lexicography and Natural Language
    Processing, pp. 37–102. John Benjamin Publishing Company (1996)
Mel'čuk, I.: Collocations and Lexical Functions. In: Cowie, A. (ed.) Phraseology. Theory,
    Analysis, and Applications, pp. 23–53. Oxford University Press, Oxford (1998)

Mel'čuk, I., Arbatchewsky-Jumarie, N., Iordanskaja, L.: Dictionnaire explicatif et combinatoire du français contemporain. In: Recherches Lexicosémantiques I, Les Presses de l'Université de Montréal (1984)

Mel'čuk, I., Arbatchewsky-Jumarie, N., Dagenais, L., Elnitsky, L., Iordanskaja, L., Lefebvre, M.-N., Mantha, S.: Dictionnaire explicatif et combinatoire du français contemporain. In: Recherches Lexico-Sémantiques II. Les Presses de l'Université de Montréal (1988)

Mel'čuk, I.: Cours de morphologie générale, vol. 1-5. Les Presses de l'Université de Montréal, CNRS Éditions, Montréal, Paris (1993–2000)

Mel'čuk, I.A., Zholkovskij, A.K.: An Explanatory Combinatorial Dictionary of the Contemporary Russian Language. Wiener Slawistischer Almanach, Sonderband 14 (1984)

Mel'čuk, I.A.: Explanatory Combinatorial Dictionary. In: Sica, G. (ed.) Open Problems in Linguistics and Lexicography, pp. 225–355. Polimetrica Publisher, Monza (2006)

Meyer, C.F.: Introducing English Linguistics. Cambridge University Press, Cambridge (2009)

Miller, G.A.: Foreword. In: Fellbaum, C. (ed.) WordNet. An Electronic Lexical Database, pp. xv–xxii. MIT Press, Cambridge (1998)

Mitchell, T.M.: The Discipline of Machine Learning. School of Computer Science, Carnegie Mellon University, Pittsburg, PA (2006)

Nastase, V., Szpakowicz, S.: Exploring noun-modifier semantic relations. In: Fifth International Workshop on Computational Semantics (IWCS-5), Tilburg, The Netherlands, pp. 285–301 (2003)

Nastase, V., Sayyad-Shiarabad, J., Sokolova, M., Szpakowicz, S.: Learning noun-modifier semantic relations with corpus-based and wordnet-based features. In: Proceedings of the Twenty-First National Conference on Artificial Intelligence and the Eighteenth Innovative Applications of Artificial Intelligence Conference. AAAI Press (2006)

Navarro, E., Sajous, F., Gaume, B., Prévot, L., Hsieh, S., Kuo, I., Magistry, P., Huang, C.R.: Wiktionary for natural language processing: methodology and limitations. In: Proceedings of the ACL 2009 Workshop, The People's Web Meets NLP: Collaboratively Constructed Semantic Resources, Suntec, Singapore, pp. 19–27 (2009)

Partridge, E.: Usage and Abusage, A Guide to Good English, 1st edn. W. W. Norton & Company, New York (1994); New edition edited by Whitcut, J.

Pecina, P.: An extensive empirical study of collocation extraction methods. In: Proceedings of the ACL 2005 Student Research Workshop, Ann Arbor, MI, pp. 13–18 (2005)

Payne, T.E.: Exploring Language Structure: A Student's Guide. Cambridge University Press, Cambridge (2006)

Polguère, A.: Towards a Theoretically-Motivated General Public Dictionary of Semantic Derivations and Collocations for French. In: Heid, U., Evert, S., Lehmann, E., Rohrer, C. (eds.) Proceedings of 9th Euralex International Congress EURALEX, Stuttgart, pp. 517–527 (2000)

Polguère, A.: Lessons from the Lexique actif du français. In: MTT 2007, Klagenfurt, May 21-24, vol. 69, Wiener Slawistischer Almanach, Sonderband (2007)

Porzig, W.: Wesenhafte Bedeutungsbeziehungen. Beträge zur Geschichte der deutsche Sprache und Literatur (58) (1934)

Princeton University "About WordNet". Princeton University (2010)
http://wordnet.princeton.edu

Roy, S.: Nearest neighbor with generalization, Christchurch, NZ (2002)

Ruppenhofer, J., Ellsworth, M., Petruck, M., Johnson, C.R., Scheffczyk, J.: FrameNet II: Extended Theory and Practice. ICSI, Berkeley (2006),
http://framenet.icsi.berkeley.edu/book/book.pdf

Rundell, M.: Macmillan Collocations Dictionary. Macmillan Publishers Limited (2010)

Sajous, F., Navarro, E., Gaume, B., Prévot, L., Chudy, Y.: Semi-automatic Endogenous Enrichment of Collaboratively Constructed Lexical Resources: Piggybacking onto Wiktionary. In: Loftsson, H., Rögnvaldsson, E., Helgadóttir, S. (eds.) IceTAL 2010. LNCS, vol. 6233, pp. 332–344. Springer, Heidelberg (2010)

Sinclair, J., Jones, S., Daley, R.: English Collocation Studies: The OSTI Report. Continuum, Antony Rowe Ltd., London, Chippenham (2004)

Steel, J. (ed.): Meaning – Text Theory. Linguistics, lexicography, and implications. University of Ottawa press (1990)

Uchida, H., Zhu, M., Senta, T.D.: Universal Networking Language. UNDL Foundation (2006)

Van Roey, J.: French-English Contrastive Lexicology: An Introduction. Peeters, Louvainla-Neuve (1990)

Vincze, O., Mosqueira, E., Alonso Ramos, M.: An online collocation dictionary of Spanish. In: Boguslavsky, I., Wanner, L. (eds.) Proceedings of the 5th International Conference on Meaning-Text Theory, Barcelona, September 8-9, pp. 275–286 (2011)

Vossen, P.: EuroWordNet: A Multilingual Database with Lexical Semantic Networks. Kluwer Academic, Dordrecht (1998)

Wanner, L.: Lexical Functions in Lexicography and Natural Language Processing. John Benjamin Publishing Company (1996)

Wanner, L.: Towards automatic fine-grained semantic classification of verb-noun collocations. Natural Language Engineering 10(2), 95–143 (2004)

Wanner, L., Bohnet, B., Giereth, M.: What is beyond Collocations? Insights from Machine Learning Experiments. In: EURALEX (2006)

Weinreich, U.: Problems in the Analysis of Idioms. In: Puhvel, J. (ed.) Substance and Structure of Language, pp. 23–82. University of California Press, CA (1969)

Weisler, S.E., Milekic, S.: Theory of Language. The MIT Press, Cambridge (2000)

Witten, I.H., Frank, E.: Data Mining: Practical machine learning tools and techniques, 2nd edn. Morgan Kaufmann, San Francisco (2005)

Zhong, Z., Tou Ng, H.: It Makes Sense: A Wide-Coverage Word Sense Disambiguation System for Free Text. In Proceedings of System Demonstrations, 48th Annual Meeting of the Association for Computational Linguistics, pp. 78–83. Uppsala University, Sweden (2010)

Appendix 1
Experimental Results

Tables presented here show the results of performance of 68 machine learning algorithms on nine training sets, one set for each of eight lexical functions chosen for out experiments, and one set for free word combinations. The presetermined of lexical functions are described and exemplified in Section 5.1.

In these tables, P stands for precision, R means recall, and F stands for F-measure calculated according to the formula given in Section 5.2. All algorithms in each table are ranked by F-measure.

Table 1 Algorithm performance for **Oper$_1$**

Algorithm	P	R	F
bayes.BayesianLogisticRegression	0.879	0.866	0.873
trees.Id3	0.879	0.861	0.870
functions.SMO	0.862	0.866	0.864
trees.FT	0.858	0.866	0.862
trees.LADTree	0.873	0.851	0.862
trees.SimpleCart	0.873	0.851	0.862
functions.SimpleLogistic	0.872	0.847	0.859
meta.ThresholdSelector	0.835	0.876	0.855
meta.EnsembleSelection	0.871	0.837	0.854
trees.BFTree	0.871	0.837	0.854
trees.ADTree	0.859	0.847	0.853
rules.JRip	0.859	0.842	0.850
meta.AttributeSelectedClassifier	0.851	0.847	0.849
meta.LogitBoost	0.862	0.837	0.849
meta.Bagging	0.854	0.842	0.848
functions.Logistic	0.787	0.916	0.847
meta.MultiClassClassifier	0.787	0.916	0.847
meta.END	0.842	0.847	0.844
meta.FilteredClassifier	0.842	0.847	0.844
meta.OrdinalClassClassifier	0.842	0.847	0.844
rules.PART	0.857	0.832	0.844
trees.J48	0.842	0.847	0.844
trees.J48graft	0.841	0.837	0.839
rules.DecisionTable	0.854	0.812	0.832

Table 1 (*continued*)

Algorithm	P	R	F
trees.REPTree	0.832	0.832	0.832
meta.RotationForest	0.850	0.812	0.830
meta.ClassificationViaRegression	0.804	0.832	0.818
rules.NNge	0.794	0.842	0.817
rules.Ridor	0.827	0.807	0.817
meta.Decorate	0.781	0.847	0.812
functions.VotedPerceptron	0.856	0.767	0.809
meta.Dagging	0.801	0.817	0.809
meta.RandomCommittee	0.742	0.827	0.782
rules.Prism	0.735	0.832	0.781
trees.RandomForest	0.735	0.822	0.776
meta.RandomSubSpace	0.913	0.624	0.741
misc.VFI	0.764	0.688	0.724
bayes.HNB	0.841	0.629	0.720
bayes.BayesNet	0.694	0.743	0.718
bayes.WAODE	0.734	0.698	0.716
bayes.AODE	0.757	0.678	0.715
bayes.NaiveBayes	0.742	0.683	0.711
bayes.NaiveBayesSimple	0.742	0.683	0.711
bayes.NaiveBayesUpdateable	0.742	0.683	0.711
trees.RandomTree	0.662	0.718	0.689
functions.RBFNetwork	0.758	0.619	0.681
lazy.LWL	0.811	0.574	0.672
bayes.AODEsr	0.691	0.599	0.642
misc.HyperPipes	0.583	0.693	0.633
lazy.IB1	0.540	0.728	0.620
lazy.IBk	0.519	0.757	0.616
lazy.KStar	0.556	0.683	0.613
meta.AdaBoostM1	0.914	0.366	0.523
functions.Winnow	0.450	0.401	0.424
meta.MultiBoostAB	0.976	0.203	0.336
rules.OneR	0.976	0.203	0.336
trees.DecisionStump	0.976	0.203	0.336
rules.ConjunctiveRule	0.857	0.208	0.335
meta.ClassificationViaClustering	0.314	0.134	0.188
functions.LibSVM	0	0	0
meta.CVParameterSelection	0	0	0
meta.Grading	0	0	0
meta.MultiScheme	0	0	0
meta.RacedIncrementalLogitBoost	0	0	0
meta.Stacking	0	0	0
meta.StackingC	0	0	0
meta.Vote	0	0	0
rules.ZeroR	0	0	0

Table 2 Algorithm performance for **CausFunc$_0$**

Algorithm	P	R	F
rules.JRip	0.747	0.705	0.725
meta.EnsembleSelection	0.744	0.659	0.699
trees.REPTree	0.750	0.648	0.695
meta.ClassificationViaRegression	0.693	0.693	0.693
trees.SimpleCart	0.727	0.636	0.679
trees.LADTree	0.674	0.682	0.678
trees.BFTree	0.733	0.625	0.675
meta.Bagging	0.726	0.602	0.658
functions.SMO	0.675	0.614	0.643
trees.ADTree	0.718	0.580	0.642
trees.FT	0.655	0.625	0.640
rules.Ridor	0.694	0.568	0.625
meta.AttributeSelectedClassifier	0.746	0.534	0.623
trees.Id3	0.618	0.625	0.621
meta.RotationForest	0.712	0.534	0.610
meta.END	0.730	0.523	0.609
meta.FilteredClassifier	0.730	0.523	0.609
meta.OrdinalClassClassifier	0.730	0.523	0.609
trees.J48	0.730	0.523	0.609
functions.Logistic	0.552	0.659	0.601
meta.MultiClassClassifier	0.552	0.659	0.601
rules.DecisionTable	0.653	0.557	0.601
rules.Prism	0.577	0.612	0.594
bayes.BayesianLogisticRegression	0.662	0.534	0.591
meta.Decorate	0.602	0.568	0.585
functions.SimpleLogistic	0.672	0.511	0.581
rules.PART	0.605	0.557	0.580
meta.RandomCommittee	0.630	0.523	0.571
trees.RandomForest	0.677	0.477	0.560
meta.LogitBoost	0.702	0.455	0.552
rules.NNge	0.595	0.500	0.543
meta.ThresholdSelector	0.597	0.489	0.538
trees.J48graft	0.745	0.398	0.519
bayes.BayesNet	0.537	0.500	0.518
misc.VFI	0.443	0.580	0.502
functions.VotedPerceptron	0.614	0.398	0.483
meta.Dagging	0.784	0.330	0.464
misc.HyperPipes	0.395	0.557	0.462
bayes.WAODE	0.690	0.330	0.446
trees.RandomTree	0.507	0.398	0.446
lazy.KStar	0.485	0.364	0.416
bayes.AODEsr	0.365	0.477	0.414
lazy.IBk	0.452	0.375	0.410

Table 2 (*continued*)

Algorithm	P	R	F
meta.RandomSubSpace	0.735	0.284	0.410
lazy.IB1	0.468	0.330	0.387
rules.OneR	0.697	0.261	0.380
bayes.NaiveBayes	0.621	0.205	0.308
bayes.NaiveBayesSimple	0.621	0.205	0.308
bayes.NaiveBayesUpdateable	0.621	0.205	0.308
bayes.AODE	0.824	0.159	0.267
functions.Winnow	0.217	0.341	0.265
functions.RBFNetwork	0.625	0.114	0.192
bayes.HNB	0.571	0.091	0.157
meta.ClassificationViaClustering	0.092	0.091	0.091
meta.AdaBoostM1	0.500	0.011	0.022
functions.LibSVM	0	0	0
lazy.LWL	0	0	0
meta.CVParameterSelection	0	0	0
meta.Grading	0	0	0
meta.MultiBoostAB	0	0	0
meta.MultiScheme	0	0	0
meta.RacedIncrementalLogitBoost	0	0	0
meta.Stacking	0	0	0
meta.StackingC	0	0	0
meta.Vote	0	0	0
rules.ConjunctiveRule	0	0	0
rules.ZeroR	0	0	0
trees.DecisionStump	0	0	0

Table 3 Algorithm performance for **CausFunc$_1$**

Algorithm	P	R	F
trees.J48	0.842	0.696	0.762
meta.FilteredClassifier	0.842	0.696	0.762
meta.END	0.842	0.696	0.762
meta.OrdinalClassClassifier	0.842	0.696	0.762
functions.SimpleLogistic	0.828	0.696	0.756
meta.LogitBoost	0.828	0.696	0.756
meta.AttributeSelectedClassifier	0.774	0.696	0.733
rules.DecisionTable	0.833	0.652	0.732
rules.JRip	0.783	0.681	0.729
trees.FT	0.742	0.710	0.726
trees.LADTree	0.818	0.652	0.726
trees.SimpleCart	0.804	0.652	0.72
trees.BFTree	0.811	0.623	0.705

Table 3 (*continued*)

Algorithm	P	R	F
trees.Id3	0.700	0.710	0.705
rules.NNge	0.742	0.667	0.702
rules.PART	0.696	0.696	0.696
trees.REPTree	0.808	0.609	0.694
trees.J48graft	0.851	0.580	0.690
meta.Bagging	0.792	0.609	0.689
bayes.BayesianLogisticRegression	0.726	0.652	0.687
meta.RotationForest	0.833	0.58	0.684
rules.Ridor	0.804	0.594	0.683
functions.SMO	0.697	0.667	0.681
meta.Dagging	0.800	0.580	0.672
functions.VotedPerceptron	0.769	0.580	0.661
lazy.LWL	0.796	0.565	0.661
meta.AdaBoostM1	0.796	0.565	0.661
meta.MultiBoostAB	0.796	0.565	0.661
rules.ConjunctiveRule	0.796	0.565	0.661
rules.OneR	0.796	0.565	0.661
trees.ADTree	0.796	0.565	0.661
trees.DecisionStump	0.796	0.565	0.661
meta.EnsembleSelection	0.78	0.565	0.655
rules.Prism	0.612	0.695	0.651
meta.ClassificationViaRegression	0.741	0.580	0.650
meta.Decorate	0.571	0.696	0.627
trees.RandomForest	0.720	0.522	0.605
meta.RandomCommittee	0.615	0.580	0.597
lazy.IBk	0.500	0.493	0.496
trees.RandomTree	0.452	0.478	0.465
lazy.IB1	0.492	0.435	0.462
bayes.WAODE	0.632	0.348	0.449
lazy.KStar	0.483	0.406	0.441
meta.ThresholdSelector	0.329	0.667	0.440
meta.RandomSubSpace	0.778	0.304	0.438
functions.Logistic	0.306	0.696	0.425
meta.MultiClassClassifier	0.306	0.696	0.425
bayes.BayesNet	0.438	0.406	0.421
bayes.AODEsr	0.375	0.478	0.42
misc.VFI	0.343	0.507	0.409
misc.HyperPipes	0.286	0.522	0.369
functions.RBFNetwork	0.462	0.174	0.253
functions.Winnow	0.163	0.348	0.222
meta.ClassificationViaClustering	0.081	0.087	0.084
bayes.AODE	0.429	0.043	0.079
bayes.HNB	0.429	0.043	0.079

Table 3 (*continued*)

Algorithm	P	R	F
bayes.NaiveBayes	0.333	0.043	0.077
bayes.NaiveBayesSimple	0.333	0.043	0.077
bayes.NaiveBayesUpdateable	0.333	0.043	0.077
functions.LibSVM	0	0	0
meta.CVParameterSelection	0	0	0
meta.Grading	0	0	0
meta.MultiScheme	0	0	0
meta.RacedIncrementalLogitBoost	0	0	0
meta.Stacking	0	0	0
meta.StackingC	0	0	0
meta.Vote	0	0	0
rules.ZeroR	0	0	0

Table 4 Algorithm performance for **Real$_1$**

Algorithm	P	R	F
rules.Prism	0.735	0.832	0.781
bayes.BayesianLogisticRegression	0.788	0.553	0.650
functions.SMO	0.722	0.553	0.627
trees.FT	0.650	0.553	0.598
rules.NNge	0.614	0.574	0.593
trees.Id3	0.600	0.574	0.587
trees.LADTree	0.733	0.468	0.571
meta.LogitBoost	0.833	0.426	0.563
rules.DecisionTable	0.750	0.447	0.560
rules.JRip	0.750	0.447	0.560
functions.SimpleLogistic	0.864	0.404	0.551
trees.J48graft	0.864	0.404	0.551
trees.BFTree	0.769	0.426	0.548
functions.Logistic	0.440	0.702	0.541
meta.MultiClassClassifier	0.440	0.702	0.541
rules.PART	0.677	0.447	0.538
meta.END	0.714	0.426	0.533
meta.FilteredClassifier	0.714	0.426	0.533
meta.OrdinalClassClassifier	0.714	0.426	0.533
trees.J48	0.714	0.426	0.533
trees.SimpleCart	0.714	0.426	0.533
meta.Bagging	0.857	0.383	0.529
misc.VFI	0.473	0.553	0.510
trees.ADTree	0.679	0.404	0.507
meta.AttributeSelectedClassifier	0.588	0.426	0.494
meta.RandomCommittee	0.613	0.404	0.487

Table 4 (*continued*)

Algorithm	P	R	F
meta.Decorate	0.453	0.511	0.480
meta.ThresholdSelector	0.409	0.000	0.478
trees.RandomForest	0.600	0.383	0.468
misc.HyperPipes	0.488	0.447	0.467
meta.EnsembleSelection	0.727	0.340	0.464
trees.REPTree	0.667	0.340	0.451
meta.RotationForest	0.778	0.298	0.431
meta.ClassificationViaRegression	0.486	0.362	0.415
rules.Ridor	0.813	0.277	0.413
lazy.LWL	0.636	0.298	0.406
meta.AdaBoostM1	0.636	0.298	0.406
meta.MultiBoostAB	0.636	0.298	0.406
rules.ConjunctiveRule	0.636	0.298	0.406
trees.DecisionStump	0.636	0.298	0.406
lazy.KStar	0.436	0.362	0.395
lazy.IBk	0.367	0.383	0.375
trees.RandomTree	0.367	0.383	0.375
lazy.IB1	0.390	0.340	0.364
meta.RandomSubSpace	0.889	0.170	0.286
bayes.BayesNet	0.357	0.213	0.267
functions.Winnow	0.211	0.340	0.260
bayes.AODEsr	0.211	0.319	0.254
functions.VotedPerceptron	0.467	0.149	0.226
rules.OneR	0.500	0.128	0.203
bayes.WAODE	0.571	0.085	0.148
meta.ClassificationViaClustering	0.071	0.106	0.085
bayes.NaiveBayes	0.333	0.021	0.040
bayes.NaiveBayesSimple	0.333	0.021	0.040
bayes.NaiveBayesUpdateable	0.333	0.021	0.040
bayes.AODE	0	0	0
bayes.HNB	0	0	0
functions.LibSVM	0	0	0
functions.RBFNetwork	0	0	0
meta.CVParameterSelection	0	0	0
meta.Dagging	0	0	0
meta.Grading	0	0	0
meta.MultiScheme	0	0	0
meta.RacedIncrementalLogitBoost	0	0	0
meta.Stacking	0	0	0
meta.StackingC	0	0	0
meta.Vote	0	0	0
rules.ZeroR	0	0	0

Table 5 Algorithm performance for **$Func_0$**

Algorithm	P	R	F
trees.BFTree	0.667	0.727	0.696
trees.Id3	0.571	0.727	0.640
meta.AttributeSelectedClassifier	0.636	0.636	0.636
meta.END	0.636	0.636	0.636
meta.FilteredClassifier	0.636	0.636	0.636
meta.OrdinalClassClassifier	0.636	0.636	0.636
misc.HyperPipes	0.636	0.636	0.636
trees.J48	0.636	0.636	0.636
lazy.LWL	0.750	0.545	0.632
meta.AdaBoostM1	0.750	0.545	0.632
meta.LogitBoost	0.750	0.545	0.632
rules.OneR	0.750	0.545	0.632
trees.DecisionStump	0.750	0.545	0.632
misc.VFI	0.583	0.636	0.609
bayes.BayesianLogisticRegression	0.667	0.545	0.600
meta.ClassificationViaRegression	0.667	0.545	0.600
rules.JRip	0.667	0.545	0.600
trees.ADTree	0.667	0.545	0.600
meta.MultiBoostAB	0.833	0.455	0.588
functions.SMO	0.538	0.636	0.583
trees.LADTree	0.538	0.636	0.583
rules.DecisionTable	0.600	0.545	0.571
rules.PART	0.600	0.545	0.571
meta.RotationForest	0.714	0.455	0.556
rules.Ridor	0.545	0.545	0.545
functions.SimpleLogistic	0.625	0.455	0.526
trees.SimpleCart	0.625	0.455	0.526
rules.NNge	0.500	0.545	0.522
trees.FT	0.667	0.364	0.471
trees.J48graft	0.667	0.364	0.471
functions.Logistic	0.368	0.636	0.467
meta.MultiClassClassifier	0.368	0.636	0.467
rules.Prism	0.364	0.571	0.444
trees.RandomTree	0.375	0.545	0.444
lazy.IBk	0.333	0.455	0.385
meta.Bagging	0.600	0.273	0.375
meta.EnsembleSelection	0.600	0.273	0.375
trees.REPTree	0.600	0.273	0.375
meta.RandomCommittee	0.500	0.273	0.353
lazy.IB1	0.333	0.364	0.348
meta.ThresholdSelector	0.308	0.364	0.333
trees.RandomForest	0.429	0.273	0.333
lazy.KStar	0.300	0.273	0.286

Table 5 (*continued*)

Algorithm	P	R	F
functions.RBFNetwork	0.500	0.182	0.267
bayes.AODEsr	0.120	0.545	0.197
functions.VotedPerceptron	1.000	0.091	0.167
meta.RandomSubSpace	1.000	0.091	0.167
rules.ConjunctiveRule	1.000	0.091	0.167
meta.Decorate	0.103	0.273	0.150
functions.Winnow	0.048	0.091	0.063
bayes.AODE	0	0	0
bayes.BayesNet	0	0	0
bayes.HNB	0	0	0
bayes.NaiveBayes	0	0	0
bayes.NaiveBayesSimple	0	0	0
bayes.NaiveBayesUpdateable	0	0	0
bayes.WAODE	0	0	0
functions.LibSVM	0	0	0
meta.ClassificationViaClustering	0	0	0
meta.CVParameterSelection	0	0	0
meta.Dagging	0	0	0
meta.Grading	0	0	0
meta.MultiScheme	0	0	0
meta.RacedIncrementalLogitBoost	0	0	0
meta.Stacking	0	0	0
meta.StackingC	0	0	0
meta.Vote	0	0	0
rules.ZeroR	0	0	0

Table 6 Algorithm performance for **Oper$_2$**

Algorithm	P	R	F
rules.PART	0.923	0.571	0.706
meta.AttributeSelectedClassifier	0.923	0.571	0.706
meta.END	0.923	0.571	0.706
meta.FilteredClassifier	0.923	0.571	0.706
meta.OrdinalClassClassifier	0.923	0.571	0.706
trees.J48	0.923	0.571	0.706
meta.LogitBoost	0.857	0.571	0.686
meta.RandomCommittee	0.722	0.619	0.667
trees.LADTree	0.667	0.667	0.667
rules.JRip	0.786	0.524	0.629
trees.SimpleCart	0.909	0.476	0.625
trees.FT	0.667	0.571	0.615
trees.BFTree	0.733	0.524	0.611

Table 6 (*continued*)

Algorithm	P	R	F
bayes.BayesianLogisticRegression	0.632	0.571	0.600
rules.NNge	0.632	0.571	0.600
functions.SMO	0.688	0.524	0.595
rules.Prism	0.545	0.632	0.585
functions.SimpleLogistic	0.900	0.429	0.581
meta.EnsembleSelection	0.900	0.429	0.581
rules.DecisionTable	0.900	0.429	0.581
rules.OneR	0.900	0.429	0.581
trees.J48graft	0.900	0.429	0.581
trees.REPTree	0.900	0.429	0.581
trees.Id3	0.542	0.619	0.578
meta.ClassificationViaRegression	0.714	0.476	0.571
trees.ADTree	0.714	0.476	0.571
meta.Bagging	0.818	0.429	0.563
functions.VotedPerceptron	0.889	0.381	0.533
rules.Ridor	0.692	0.429	0.529
meta.RotationForest	0.727	0.381	0.500
lazy.KStar	0.476	0.476	0.476
trees.RandomTree	0.407	0.524	0.458
meta.RandomSubSpace	0.857	0.286	0.429
lazy.IBk	0.370	0.476	0.417
trees.RandomForest	0.538	0.333	0.412
meta.ThresholdSelector	0.333	0.524	0.407
functions.Logistic	0.302	0.619	0.406
meta.MultiClassClassifier	0.302	0.619	0.406
misc.HyperPipes	0.323	0.476	0.385
misc.VFI	0.303	0.476	0.370
bayes.AODEsr	0.245	0.571	0.343
meta.Decorate	0.256	0.476	0.333
lazy.IB1	0.286	0.381	0.327
meta.Dagging	1.000	0.190	0.320
meta.MultiBoostAB	0.667	0.095	0.167
functions.Winnow	0.067	0.143	0.091
rules.ConjunctiveRule	1.000	0.048	0.091
meta.AdaBoostM1	0.500	0.048	0.087
trees.DecisionStump	0.500	0.048	0.087
lazy.LWL	0.333	0.048	0.083
meta.ClassificationViaClustering	0.050	0.095	0.066
bayes.AODE	0	0	0
bayes.BayesNet	0	0	0
bayes.HNB	0	0	0
bayes.NaiveBayes	0	0	0
bayes.NaiveBayesSimple	0	0	0

Table 6 (*continued*)

Algorithm	P	R	F
bayes.NaiveBayesUpdateable	0	0	0
bayes.WAODE	0	0	0
functions.LibSVM	0	0	0
functions.RBFNetwork	0	0	0
meta.CVParameterSelection	0	0	0
meta.Grading	0	0	0
meta.MultiScheme	0	0	0
meta.RacedIncrementalLogitBoost	0	0	0
meta.Stacking	0	0	0
meta.StackingC	0	0	0
meta.Vote	0	0	0
rules.ZeroR	0	0	0

Table 7 Algorithm performance for **IncepOper$_1$**

Algorithm	P	R	F
rules.Prism	0.750	0.800	0.774
rules.NNge	0.923	0.600	0.727
functions.SMO	0.813	0.650	0.722
functions.SimpleLogistic	0.857	0.600	0.706
bayes.BayesianLogisticRegression	0.917	0.550	0.687
trees.LADTree	0.917	0.550	0.687
trees.Id3	0.846	0.550	0.667
trees.FT	0.786	0.550	0.647
misc.VFI	0.733	0.550	0.629
meta.LogitBoost	0.769	0.500	0.606
meta.RandomCommittee	0.818	0.450	0.581
meta.AttributeSelectedClassifier	0.667	0.500	0.571
meta.END	0.667	0.500	0.571
meta.FilteredClassifier	0.667	0.500	0.571
meta.OrdinalClassClassifier	0.667	0.500	0.571
rules.DecisionTable	0.667	0.500	0.571
trees.BFTree	0.667	0.500	0.571
trees.J48	0.667	0.500	0.571
rules.JRip	0.625	0.500	0.556
trees.RandomForest	0.889	0.400	0.552
trees.ADTree	0.692	0.450	0.545
rules.PART	0.556	0.500	0.526
lazy.LWL	0.727	0.400	0.516
trees.SimpleCart	0.727	0.400	0.516
trees.RandomTree	0.600	0.450	0.514
meta.RandomSubSpace	0.778	0.350	0.483
trees.REPTree	0.700	0.350	0.467

Table 7 (*continued*)

Algorithm	P	R	F
misc.HyperPipes	0.636	0.350	0.452
meta.Decorate	0.343	0.600	0.436
lazy.KStar	0.750	0.300	0.429
functions.Logistic	0.324	0.600	0.421
meta.MultiClassClassifier	0.324	0.600	0.421
lazy.IBk	0.600	0.300	0.400
meta.RotationForest	0.714	0.250	0.370
functions.VotedPerceptron	0.625	0.250	0.357
lazy.IB1	0.625	0.250	0.357
trees.J48graft	0.625	0.250	0.357
meta.ThresholdSelector	0.296	0.400	0.340
rules.Ridor	0.500	0.250	0.333
bayes.AODEsr	0.233	0.500	0.317
meta.ClassificationViaRegression	0.571	0.200	0.296
meta.Bagging	0.600	0.150	0.240
meta.AdaBoostM1	0.500	0.100	0.167
meta.MultiBoostAB	0.500	0.100	0.167
rules.OneR	0.500	0.100	0.167
trees.DecisionStump	0.500	0.100	0.167
meta.EnsembleSelection	0.400	0.100	0.160
rules.ConjunctiveRule	0.500	0.050	0.091
meta.ClassificationViaClustering	0.016	0.050	0.024
bayes.AODE	0	0	0
bayes.BayesNet	0	0	0
bayes.HNB	0	0	0
bayes.NaiveBayes	0	0	0
bayes.NaiveBayesSimple	0	0	0
bayes.NaiveBayesUpdateable	0	0	0
bayes.WAODE	0	0	0
functions.LibSVM	0	0	0
functions.RBFNetwork	0	0	0
functions.Winnow	0	0	0
meta.CVParameterSelection	0	0	0
meta.Dagging	0	0	0
meta.Grading	0	0	0
meta.MultiScheme	0	0	0
meta.RacedIncrementalLogitBoost	0	0	0
meta.Stacking	0	0	0
meta.StackingC	0	0	0
meta.Vote	0	0	0
rules.ZeroR	0	0	0

Table 8 Algorithm performance for **ContOper$_1$**

Algorithm	P	R	F
functions.SimpleLogistic	0.909	0.769	0.833
rules.DecisionTable	0.909	0.769	0.833
meta.AttributeSelectedClassifier	0.833	0.769	0.800
meta.END	0.833	0.769	0.800
meta.FilteredClassifier	0.833	0.769	0.800
meta.OrdinalClassClassifier	0.833	0.769	0.800
rules.JRip	0.833	0.769	0.800
rules.PART	0.833	0.769	0.800
trees.BFTree	0.833	0.769	0.800
trees.J48	0.833	0.769	0.800
trees.SimpleCart	0.833	0.769	0.800
functions.Logistic	0.733	0.846	0.786
meta.MultiClassClassifier	0.733	0.846	0.786
meta.Bagging	0.900	0.692	0.783
rules.Prism	0.750	0.818	0.783
rules.Ridor	0.90	0.692	0.783
functions.SMO	0.818	0.692	0.750
meta.LogitBoost	0.818	0.692	0.750
rules.NNge	0.818	0.692	0.750
trees.FT	0.818	0.692	0.750
trees.Id3	0.818	0.692	0.750
trees.LADTree	0.818	0.692	0.750
meta.EnsembleSelection	0.889	0.615	0.727
meta.RandomSubSpace	0.889	0.615	0.727
lazy.LWL	0.800	0.615	0.696
trees.ADTree	0.800	0.615	0.696
trees.REPTree	0.800	0.615	0.696
bayes.BayesianLogisticRegression	0.778	0.538	0.636
meta.ClassificationViaRegression	0.857	0.462	0.600
meta.MultiBoostAB	0.857	0.462	0.600
rules.OneR	0.857	0.462	0.600
trees.DecisionStump	0.857	0.462	0.600
trees.RandomForest	0.857	0.462	0.600
functions.VotedPerceptron	0.750	0.462	0.571
meta.RandomCommittee	0.750	0.462	0.571
meta.AdaBoostM1	0.833	0.385	0.526
meta.RotationForest	0.833	0.385	0.526
meta.ThresholdSelector	0.714	0.385	0.500
trees.J48graft	0.714	0.385	0.500
lazy.IBk	0.625	0.385	0.476
meta.Decorate	0.345	0.769	0.476
trees.RandomTree	0.556	0.385	0.455
lazy.KStar	0.667	0.308	0.421

Table 8 (*continued*)

Algorithm	P	R	F
rules.ConjunctiveRule	1.000	0.231	0.375
lazy.IB1	0.600	0.231	0.333
misc.VFI	0.227	0.385	0.286
bayes.AODEsr	0.143	0.538	0.226
misc.HyperPipes	0.150	0.231	0.182
meta.Dagging	0.500	0.077	0.133
functions.RBFNetwork	0.333	0.077	0.125
functions.Winnow	0.095	0.154	0.118
bayes.AODE	0	0	0
bayes.BayesNet	0	0	0
bayes.HNB	0	0	0
bayes.NaiveBayes	0	0	0
bayes.NaiveBayesSimple	0	0	0
bayes.NaiveBayesUpdateable	0	0	0
bayes.WAODE	0	0	0
functions.LibSVM	0	0	0
meta.ClassificationViaClustering	0	0	0
meta.CVParameterSelection	0	0	0
meta.Grading	0	0	0
meta.MultiScheme	0	0	0
meta.RacedIncrementalLogitBoost	0	0	0
meta.Stacking	0	0	0
meta.StackingC	0	0	0
meta.Vote	0	0	0
rules.ZeroR	0	0	0

Table 9 Algorithm performance for FWC

Algorithm	P	R	F
rules.Prism	0.639	0.702	0.669
bayes.BayesianLogisticRegression	0.658	0.629	0.643
functions.SMO	0.656	0.623	0.639
bayes.BayesNet	0.609	0.667	0.637
meta.RandomCommittee	0.639	0.635	0.637
trees.Id3	0.627	0.635	0.631
trees.FT	0.642	0.610	0.626
meta.Decorate	0.627	0.591	0.608
misc.VFI	0.542	0.692	0.608
rules.NNge	0.628	0.585	0.606
bayes.WAODE	0.634	0.579	0.605
lazy.IBk	0.577	0.635	0.605
meta.RotationForest	0.701	0.516	0.594
trees.RandomForest	0.625	0.566	0.594

Table 9 (*continued*)

Algorithm	P	R	F
trees.RandomTree	0.611	0.572	0.591
trees.LADTree	0.608	0.566	0.586
lazy.KStar	0.599	0.572	0.585
lazy.IB1	0.571	0.585	0.578
meta.EnsembleSelection	0.731	0.478	0.578
trees.ADTree	0.636	0.528	0.577
meta.Bagging	0.700	0.484	0.572
trees.REPTree	0.607	0.535	0.569
functions.SimpleLogistic	0.718	0.465	0.565
meta.ClassificationViaRegression	0.612	0.516	0.560
trees.BFTree	0.688	0.472	0.560
bayes.AODEsr	0.525	0.597	0.559
rules.PART	0.609	0.509	0.555
functions.VotedPerceptron	0.627	0.497	0.554
trees.SimpleCart	0.622	0.497	0.552
bayes.NaiveBayes	0.626	0.484	0.546
bayes.NaiveBayesSimple	0.626	0.484	0.546
bayes.NaiveBayesUpdateable	0.626	0.484	0.546
rules.DecisionTable	0.626	0.484	0.546
bayes.AODE	0.647	0.472	0.545
misc.HyperPipes	0.440	0.711	0.543
meta.Dagging	0.719	0.434	0.541
rules.JRip	0.610	0.472	0.532
functions.Logistic	0.449	0.560	0.499
meta.MultiClassClassifier	0.449	0.560	0.499
meta.ThresholdSelector	0.435	0.585	0.499
functions.RBFNetwork	0.625	0.409	0.494
meta.END	0.714	0.377	0.494
meta.FilteredClassifier	0.714	0.377	0.494
meta.OrdinalClassClassifier	0.714	0.377	0.494
meta.RandomSubSpace	0.714	0.377	0.494
trees.J48	0.714	0.377	0.494
meta.AttributeSelectedClassifier	0.687	0.358	0.471
trees.J48graft	0.746	0.333	0.461
lazy.LWL	0.557	0.371	0.445
bayes.HNB	0.718	0.321	0.443
rules.Ridor	0.654	0.321	0.430
meta.AdaBoostM1	0.581	0.340	0.429
trees.DecisionStump	0.550	0.346	0.425
meta.MultiBoostAB	0.552	0.333	0.416
functions.Winnow	0.374	0.447	0.407
meta.LogitBoost	0.638	0.277	0.386
rules.ConjunctiveRule	0.551	0.239	0.333

Table 9 (*continued*)

Algorithm	P	R	F
rules.OneR	0.419	0.164	0.235
meta.ClassificationViaClustering	0.188	0.132	0.155
functions.LibSVM	0	0	0
meta.CVParameterSelection	0	0	0
meta.Grading	0	0	0
meta.MultiScheme	0	0	0
meta.RacedIncrementalLogitBoost	0	0	0
meta.Stacking	0	0	0
meta.StackingC	0	0	0
meta.Vote	0	0	0
rules.ZeroR	0	0	0

Appendix 2
Dictionary of Spanish Verbal Lexical Functions

Here we present our dictionary of Spanish verbal lexical funcitons. The lexical material is organized in a table format. The following names and abbreviations are used in the Dictionary:

- **LF:** lexical function
- **FWC:** free word combination
- **ERROR:** combination erroneously included in the list by the software and manually marked as such.
- **Verb Sense Number:** sense number for the verb in the Spanish WordNet
- **Noun Sense Number:** sense number for the noun in the Spanish WordNet
- **N/A:** no sense is available for the respective verb/noun in the Spanish WordNet such that this sense is appropriate in the context in which a given verb-noun pair is used in the Spanish Web Corpus
- **FREQ:** frequency of the respective verb-noun pair in the Spanish Web Corpus
- **Spanish WordNet:** Vossen P. (ed.) 1998. EuroWordNet: A Multilingual Database with Lexical Semantic Networks. Dordrecht: Kluwer Academic Publishers, available at http://www.lsi.upc.edu/~nlp/web/index.php? Itemid=57&id=31&option=com_content&task=view
- **Spanish Web Corpus:** Spanish Web Corpus in the Sketch Engine. 3 May 2010. http://trac.sketchengine.co.uk/wiki/Corpora/ SpanishWebCorpus

LF/ FWC/ ERROR	Verb	Verb Sense Number	Noun	Noun Sense Number	Frequency
Oper1	*dar*	2	*cuenta*	N/A	9236
CausFunc0	*formar*	2	*parte*	1	7454
Oper1	*tener*	1	*lugar*	4	6680

Oper1	*tener*	1	*derecho*	1	5255
CausFunc1	*hacer*	2	*falta*	N/A	4827
CausFunc1	*dar*	9	*lugar*	4	4180
Oper1	*hacer*	15	*referencia*	2	3252
Func0	*hacer*	N/A	*año*	2	3211
Oper1	*tener*	1	*problema*	7	3075
Func0	*hacer*	N/A	*tiempo*	1	3059
IncepOper1	*tomar*	4	*decisión*	2	2781
Oper1	*tener*	1	*acceso*	3	2773
Oper1	*tener*	1	*razón*	2	2768
Caus2Func1	*llamar*	8	*atención*	1	2698
Oper1	*tener*	1	*sentido*	1	2563
ERROR	*haber*		*estado*		2430
FWC	*hacer*	6	*cosa*	3	2374
Oper1	*tener*	3	*miedo*	1	2226
ERROR	*haber*		*hecho*		2168
Oper1	*tener*	1	*oportunidad*	1	2137
CausFunc1	*dar*	9	*paso*	4	2100
Oper1	*hacer*	15	*uso*	1	2081
Real1	*resolver*	2	*problema*	7	1939
FWC	*tener*	1	*tiempo*	1	1905
Oper1	*tener*	1	*efecto*	1	1905
Oper1	*prestar*	1	*atención*	1	1883
Oper1	*tener*	1	*relación*	4	1849
Oper1	*tener*	1	*capacidad*	4	1847
Oper1	*tener*	1	*valor*	2	1769
FWC	*valer*	1	*pena*	5	1765
Oper1	*tener*	1	*idea*	5	1734
FWC	*tener*	1	*hijo*	1	1685
Oper1	*dar*	9	*vuelta*	6	1653
Oper1	*tener*	1	*posibilidad*	2	1603
Oper1	*tener*	1	*éxito*	3	1596
FWC	*abrir*	1	*puerta*	1	1591
CausFunc1	*dar*	9	*respuesta*	2	1576
FWC	*merecer*	1	*pena*	12	1521

Oper1	*tener*	2	*carácter*	1	1479
Oper1	*tener*	1	*suerte*	3	1475
Oper1	*tener*	1	*conocimiento*	3	1396
Oper1	*jugar*	4	*papel*	1	1394
Oper1	*tener*	2	*importancia*	1	1372
Oper1	*hacer*	15	*caso*	N/A	1370
PerfFunc0	*llegar*	3	*momento*	1	1293
Oper1	*tener*	1	*interés*	1	1245
CausFunc1	*poner*	4	*fin*	3	1235
CausFunc0	*publicar*	2	*comentario*	1	1232
PerfOper1	*tomar*	4	*medida*	3	1216
Oper1	*hacer*	15	*esfuerzo*	2	1202
Oper1	*correr*	3	*riesgo*	1	1159
Func1	*caber*	1	*duda*	2	1145
FWC	*decir*	2	*verdad*	3	1132
Oper1	*tener*	1	*experiencia*	1	1114
Oper1	*tener*	1	*necesidad*	1	1102
FWC	*obtener*	1	*información*	1	1091
ERROR	«		*no*		1086
CausFunc1	*dar*	9	*vida*	2	1086
Oper1	*hacer*	15	*pregunta*	1	1075
Oper1	*hacer*	15	*daño*	3	1074
IncepOper1	*iniciar*	4	*sesión*	3	1062
Oper1	*hacer*	15	*trabajo*	1	1035
Func0	*pasar*	2	*día*	1	1026
Func0	*pasar*	2	*año*	2	1000
Func0	*pasar*	2	*tiempo*	1	988
Func0	*hacer*	N/A	*día*	1	971
Oper1	*prestar*	1	*servicio*	1	951
Func0	*hacer*	N/A	*mes*	1	950
FWC	*tener*	1	*cuenta*	6	944
IncepOper1	*tomar*	4	*parte*	1	939
FWC	*enviar*	4	*mensaje*	2	929
ERROR	«		*La*		927
FWC	*decir*	2	*cosa*	5	924

ERROR	decir		no		922
Real2	leer	1	libro	1	910
Oper1	ocupar	2	lugar	4	905
CausFunc1	dar	9	importancia	1	899
Caus2Func1	dar	9	impresión	2	898
ERROR	dar\decir		cuenta		897
Oper1	tener	1	vida	2	892
Real1	solucionar	1	problema	7	884
CausFunc1	dar	9	nombre	1	874
Oper1	tener	1	dificultad	3	873
Oper1	tener	1	gana	N/A	871
Caus1Oper1	dar	9	resultado	2	871
Oper1	hacer	15	cargo	4	870
FWC	contar	2	historia	3	870
Oper1	desempeñar	N/A	papel	1	869
Oper1	dar	3	gracia	N/A	864
FWC	ganar	3	dinero	1	863
CausFunc1	dar	9	razón	2	861
Real1	satisfacer	1	necesidad	1	858
Oper1	realizar	6	estudio	5	854
CausFunc0	establecer	2	relación	6	847
Oper1	adoptar	1	medida	3	846
CausFunc0	producir	4	efecto	1	840
CausFunc1	dar	9	origen	2	840
IncepOper1	tomar	4	conciencia	1	839
Oper1	tener	1	obligación	1	839
Oper1	tener	1	consecuencia	1	833
Oper2	obtener	1	resultado	2	829
ContOper1	mantener	2	relación	4	829
FWC	perder	1	tiempo	1	827
Oper1	hacer	15	hincapié	N/A	827
Oper1	tener	1	poder	1	812
Oper1	tener	1	fuerza	2	811
ERROR	parecer		no		807
CausFunc0	escribir	1	libro	1	802

CausPlusFunc0	mejorar	1	calidad	1	784
Oper1	realizar	6	trabajo	1	779
CausFunc1	dar	9	idea	5	773
CausFunc1	hacer	2	realidad	5	770
Oper1	tener	1	forma	7	766
FWC	ver	1	luz	1	762
Oper1	tener	1	papel	1	758
CausFunc0	poner	4	ejemplo	1	758
FWC	tener	1	dinero	1	755
FWC	ignorar	1	hilo	N/A	745
Oper1	tener	1	conciencia	1	741
FWC	ver	1	cosa	3	740
Oper1	tener	1	duda	2	737
ERROR	haber		oído		733
CausFunc1	abrir	5	camino	6	732
CausFunc1	dar	9	sentido	1	730
FWC	dar	4	tiempo	1	724
Func0	pasar	2	mes	1	723
Oper1	hacer	15	comentario	1	722
Oper1	tener	1	ventaja	1	719
CausFunc1	dar	9	oportunidad	1	711
CausFunc0	tomar	1	nota	2	710
Oper1	tener	1	nombre	1	705
Oper1	tener	3	sensación	2	694
FWC	abrir	1	ojo	1	691
CausFunc1	dar	9	forma	7	689
Oper1	realizar	6	actividad	1	683
Manif	plantear	1	problema	7	682
Oper1	tener	1	información	1	678
Oper1	tener	1	intención	1	676
CausFunc1	proporcionar	1	información	1	676
Oper1	tener	1	ocasión	2	672
Real1	alcanzar	1	nivel	1	668
CausFunc1	ofrecer	1	servicio	1	666
Oper1	cometer	1	error	1	654

Oper1	tener	1	responsabilidad	1	652
ERROR	hacer		no		650
FWC	pedir	2	ayuda	1	649
FWC	ver	1	película	1	637
Oper1	echar	N/A	mano	N/A	635
ERROR	haber		sentido		633
Func0	existir	1	posibilidad	2	633
CausFunc1	producir	4	cambio	3	627
FWC	dar	4	información	1	619
CausFunc1	dar	9	clase	2	619
Oper1	tener	1	impacto	3	615
CausFunc0	encontrar	5	solución	1	609
Oper2	recibir	1	información	1	604
Oper1	pedir	2	perdón	1	603
ERROR	hacer		par		600
IncepOper1	asumir	2	responsabilidad	1	598
Caus2Func1	dar	9	miedo	1	596
Oper1	echar	1	vistazo	1	592
Oper1	tener	1	influencia	1	591
FWC	cerrar	3	ojo	1	588
FWC	tratar	3	tema	1	583
Oper1	hacer	15	amor	N/A	582
FWC	salvar	1	vida	8	574
Oper1	hacer	15	mención	1	574
Oper1	tener	1	visión	3	569
Oper1	hacer	15	eco	2	567
Oper1	tener	1	respuesta	2	564
Oper1	hacer	15	ejercicio	2	564
CausFunc1	dar	9	ejemplo	1	564
Oper1	tener	1	nivel	1	563
ERROR	hacer		frente		563
FWC	buscar	2	solución	1	563
ERROR	·		La		561
CausFunc1	causar	1	daño	3	556
CausFunc1	ofrecer	1	posibilidad	2	552

FWC	*llevar*	5	*tiempo*	1	551
CausFunc1	*ofrecer*	1	*información*	1	549
Caus2Func1	*hacer*	2	*gracia*	3	549
FWC	*ser*	1	*persona*	1	546
FWC	*enviar*	4	*comentario*	1	544
Oper1	*tener*	1	*contacto*	1	540
Oper1	*tener*	1	*significado*	2	536
Oper1	*tener*	1	*remedio*	1	535
ERROR	*tener*		*mas*		535
ERROR	*decidir*		*no*		527
Caus2Func1	*dar*	9	*sensación*	2	527
Func0	*hacer*	N/A	*semana*	1	521
Oper3	*contener*	4	*información*	1	520
CausFunc0	*publicar*	1	*libro*	1	516
ERROR	*conciliar*		*vaticano*		515
FWC	*ganar*	3	*vida*	2	513
ContOper1	*seguir*	11	*camino*	6	510
FWC	*necesitar*	2	*ayuda*	1	510
Oper1	*dar*	3	*salto*	1	509
CausFunc0	*encontrar*	6	*información*	1	507
Func0	*existir*	1	*diferencia*	4	506
CausFunc1	*abrir*	5	*paso*	4	506
Oper1	*tener*	3	*palabra*	2	505
Oper1	*tener*	3	*impresión*	2	504
IncepOper1	*recibir*	1	*nombre*	1	503
ERROR	«		»		501
ContOper1	*llevar*	5	*vida*	5	500
IncepOper1	*iniciar*	4	*proceso*	3	500
Oper1	*tener*	1	*repercusión*	1	497
Oper1	*dar*	3	*testimonio*	2	496
Oper1	*hacer*	15	*prueba*	2	493
Oper1	*dar*	3	*orden*	2	492
Oper1	*tener*	1	*función*	1	491
Real1	*cumplir*	3	*requisito*	1	488
Oper1	*tener*	1	*libertad*	1	487

FWC	*referir*	1	*artículo*	3	484
ContOper1	*vivir*	2	*vida*	5	482
Oper2	*recibir*	1	*premio*	2	477
FWC	*cerrar*	3	*puerta*	1	477
CausFunc0	*agregar*	1	*comentario*	1	477
FWC	*tener*	1	*ojo*	1	475
FWC	*llevar*	5	*año*	2	474
Oper1	*hacer*	15	*estudio*	5	471
Oper1	*tener*	1	*trabajo*	6	470
ERROR	*preferir*		*no*		469
Oper1	*hacer*	15	*análisis*	1	467
Oper1	*decir*	1	*palabra*	2	467
Real1	*leer*	1	*comentario*	1	466
CausFunc0	*dar*	9	*explicación*	1	466
CausFunc0	*crear*	1	*cuenta*	6	466
IncepOper1	*tomar*	4	*forma*	7	465
Oper1	*tener*	1	*opción*	1	463
FWC	*poner*	1	*mano*	1	461
Func0	*hacer*	N/A	*siglo*	1	459
IncepOper1	*tomar*	6	*posesión*	2	458
Oper2	*obtener*	1	*beneficio*	3	458
CausFunc1	*dar*	9	*pie*	3	458
FWC	*tener*	2	*precio*	2	457
Oper1	*hacer*	15	*justicia*	2	456
Caus2Func1	*dar*	9	*pena*	12	456
Oper1	*tener*	1	*característica*	3	454
FWC	*tener*	1	*cosa*	3	452
FWC	*buscar*	2	*información*	1	452
CausFunc1	*poner*	4	*énfasis*	2	450
Oper1	*tener*	1	*aspecto*	4	448
Oper1	*correr*	3	*peligro*	4	448
Real1	*cumplir*	3	*función*	1	447
FWC	*tener*	1	*sistema*	3	445
FWC	*tener*	1	*amigo*	1	445
Oper1	*realizar*	6	*tarea*	1	443

ERROR	»		no		442
PermOper1	*permitir*	1	*acceso*	3	439
CausPlusFunc0	*mejorar*	1	*condición*	2	437
Oper1	*tener*	1	*control*	1	435
IncepFunc0	*llegar*	3	*hora*	1	430
Oper1	*tener*	1	*fe*	1	429
Oper1	*hacer*	15	*negocio*	2	429
CausFunc0	*escribir*	1	*artículo*	3	429
Real2	*ganar*	2	*elección*	1	428
CausFunc0	*desarrollar*	2	*actividad*	1	428
Oper1	*hacer*	15	*click*	N/A	426
ERROR	*hacer*		*mas*		423
Oper1	*tener*	1	*parte*	1	422
FWC	*pasar*	2	*noche*	1	422
ERROR	*dar*		*Vinci*		422
ERROR	*hacer*		*tus*		421
CausFunc1	*reservar*	4	*derecho*	1	419
IncepOper1	*encontrar*	7	*trabajo*	6	417
Oper1	*tener*	1	*cabida*	1	416
CausFunc1	*dar*	9	*prioridad*	1	414
ERROR	*ver*		*no*		413
Oper1	*hacer*	15	*vida*	2	413
CausPlusFunc1	*facilitar*	1	*acceso*	3	412
FWC	*escuchar*	1	*música*	1	412
CausFunc0	*escribir*	1	*carta*	1	412
CausFunc0	*convocar*	1	*concurso*	1	412
IncepFunc0	*llegar*	3	*día*	2	411
FWC	*respetar*	1	*derecho*	1	410
CausFunc0	*publicar*	1	*artículo*	3	410
Real1	*leer*	1	*artículo*	3	410
CausFunc0	*encontrar*	5	*respuesta*	2	409
Caus1Oper1	*contraer*	4	*matrimonio*	1	409
Oper1	*tener*	3	*confianza*	2	408
CausManifFunc0	*plantear*	1	*cuestión*	2	407
Oper1	*tener*	1	*origen*	2	406

Oper1	*tener*	1	*base*	1	404
FWC	*celebrar*	1	*día*	2	404
Oper1	*realizar*	6	*análisis*	1	403
Oper1	*tener*	3	*hambre*	2	402
CausFunc1	*dar*	9	*caso*	2	397
CausFunc1	*poner*	4	*nombre*	1	396
Real1	*alcanzar*	1	*objetivo*	6	396
Oper1	*tener*	2	*peso*	3	395
Oper1	*pulsar*	1	*botón*	2	395
Copul	*ser*	1	*hombre*	3	394
Oper1	*tener*	3	*esperanza*	3	393
Caus1Func1	*hacer*	2	*parte*	1	389
CausFunc1	*dar*	9	*muestra*	9	387
CausFunc0	*desarrollar*	2	*programa*	1	386
Oper1	*pedir*	2	*disculpa*	1	385
Oper1	*hacer*	15	*cambio*	3	384
Func0	*pasar*	2	*hora*	1	383
Oper1	*tener*	1	*medio*	3	382
FWC	*defender*	3	*derecho*	1	382
CausFunc1	*crear*	1	*condición*	5	382
Real3	*reconocer*	2	*derecho*	1	381
CausManifFunc0	*anunciar*	1	*concurso*	1	381
CausFunc0	*dejar*	17	*constancia*	1	379
CausFunc0	*hacer*	2	*hombre*	3	376
Oper1	*tener*	3	*paciencia*	1	375
Oper1	*presentar*	1	*problema*	7	375
ContOper1	*seguir*	11	*paso*	4	374
FWC	*personalizar*	1	*sitio*	1	374
ERROR	*pasar*		*no*		374
Oper1	*tener*	1	*seguridad*	3	373
Oper1	*tener*	1	*presencia*	2	373
ERROR	*tener*		*no*		373
Oper1	*pedir*	2	*permiso*	1	371
Oper1	*hacer*	15	*favor*	1	371
Func0	*pasar*	2	*rato*	N/A	370

Real1	*ganar*	5	*premio*	2	370
PlusOper1	*acumular*	1	*reputación*	3	370
FWC	*dedicar*	1	*tiempo*	1	369
FWC	*buscar*	2	*trabajo*	6	369
ERROR	*haber*		*leido*		368
Oper1	*tener*	1	*solución*	1	366
Oper1	*tener*	1	*límite*	1	366
CausFunc0	*firmar*	4	*acuerdo*	3	366
FWC	*cambiar*	1	*mundo*	2	366
FWC	*abordar*	2	*tema*	1	363
CausFunc0	*dejar*	4	*huella*	1	362
Oper1	*tener*	3	*consideración*	2	360
Oper2	*sacar*	4	*partido*	N/A	359
Func0	*existir*	1	*relación*	4	359
CausPlusFunc1	*aumentar*	2	*número*	1	359
FinOper1	*perder*	1	*vida*	2	358
CausFunc0	*hacer*	2	*idea*	5	356
Oper1	*realizar*	6	*esfuerzo*	2	354
Oper1	*tener*	1	*historia*	2	353
CausFunc0	*hacer*	2	*foto*	1	353
Caus1Func1	*sacar*	4	*provecho*	3	352
CausFunc1	*dar*	9	*solución*	1	352
Oper1	*tener*	1	*duración*	4	351
CausMinusFunc0	*reducir*	2	*riesgo*	1	351
CausFunc1	*dar*	9	*fe*	1	351
CausFunc1	*dar*	9	*comienzo*	2	351
Oper1	*tener*	1	*fin*	3	350
FWC	*desear*	1	*Hostalia*	N/A	350
CausFunc1	*dar*	9	*acceso*	3	350
Oper1	*tener*	1	*voz*	2	346
FWC	*tener*	1	*número*	2	345
Real1	*pagar*	1	*precio*	2	345
Caus1Func1	*sacar*	4	*conclusión*	4	344
Func0	*hacer*	N/A	*década*	1	344
CausFunc0	*formar*	1	*grupo*	1	343

CausFunc1	*dar*	9	*posibilidad*	2	343
Oper1	*realizar*	6	*investigación*	3	341
Oper1	*realizar*	6	*acción*	1	341
CausFunc1	*dar*	9	*fuerza*	2	341
Oper2	*recibir*	1	*mensaje*	2	340
CausFunc0	*hacer*	2	*película*	1	340
ERROR	*esperar*		*no*		340
CausFunc1	*dar*	9	*valor*	2	340
Oper1	*dar*	9	*bienvenida*	2	340
FWC	*ver*	1	*mundo*	2	339
Oper1	*tener*	1	*objetivo*	6	338
ContOper1	*guardar*	1	*relación*	4	338
CausFunc1	*dar*	4	*servicio*	1	338
Oper2	*oír*	1	*voz*	2	337
CausFunc0	*hacer*	2	*obra*	1	337
Oper1	*tener*	1	*recurso*	2	335
CausFunc0	*establecer*	2	*sistema*	3	332
Oper1	*tener*	1	*espacio*	2	331
ContOper1	*mantener*	2	*contacto*	1	330
CausFunc0	*hacer*	2	*guerra*	1	330
CausFunc0	*garantizar*	3	*seguridad*	2	330
Oper1	*ocupar*	2	*espacio*	2	329
Oper1	*ejercer*	2	*influencia*	1	329
CausFunc0	*desempeñar*	N/A	*función*	1	329
Caus2Func1	*dar*	9	*gana*	N/A	329
Real1	*utilizar*	1	*sistema*	3	327
Oper1	*tener*	1	*probabilidad*	2	327
Oper1	*tener*	1	*tendencia*	5	325
Oper1	*ocupar*	3	*cargo*	2	325
Oper1	*llevar*	8	*nombre*	1	325
CausFunc0	*hacer*	2	*política*	1	325
Oper2	*recibir*	1	*tratamiento*	1	324
CausFunc1	*ofrecer*	1	*oportunidad*	1	324
CausFunc0	*interponer*	2	*recurso*	4	324
CausFunc0	*dar*	9	*luz*	1	324

Oper1	*realizar*	6	*búsqueda*	1	323
Oper1	*tener*	1	*programa*	1	322
Oper1	*rendir*	2	*homenaje*	1	322
Oper1	*tener*	1	*propiedad*	3	321
CausFunc0	*hacer*	2	*ruido*	1	321
Real1	*seguir*	11	*instrucción*	1	320
Oper1	*realizar*	6	*prueba*	2	320
Real1	*leer*	1	*texto*	1	320
Real1	*aprobar*	1	*ley*	3	319
CausFunc0	*dar*	9	*fruto*	2	318
Oper1	*dar*	3	*golpe*	3	317
Oper1	*tener*	1	*resultado*	2	316
Oper1	*tener*	1	*punto*	16	316
Real1	*ver*	1	*televisión*	1	315
Real1	*beber*	1	*agua*	1	315
Oper2	*adquirir*	2	*conocimiento*	3	315
Real1	*leer*	1	*opinión*	2	313
CausFunc0	*hacer*	2	*campaña*	1	313
Copul	*parecer*	1	*mentira*	1	311
Oper1	*hacer*	15	*clic*	1	310
FWC	*abrir*	1	*ventana*	1	310
Func0	*hacer*	N/A	*rato*	N/A	309
CausFunc0	*encontrar*	6	*forma*	7	309
CausFunc0	*comprar*	1	*libro*	1	309
Copul	*ser*	1	*hijo*	1	308
IncepOper1	*tomar*	4	*iniciativa*	1	307
Oper1	*tomar*	5	*café*	1	307
Oper1	*hacer*	15	*declaración*	1	307
Oper1	*hacer*	15	*alusión*	1	306
Real1	*cubrir*	6	*necesidad*	1	306
CausFunc1	*constituir*	2	*base*	3	306
ERROR	—		*no*		305
CausFunc1	*dar*	4	*consejo*	2	305
Oper1	*tener*	1	*riesgo*	1	304
Real1	*cumplir*	3	*ley*	3	304

FWC	cambiar	1	cosa	3	304
FWC	enviar	4	correo	2	303
FWC	decir	1	don	3	303
Real1	usar	1	palabra	1	302
Oper1	poner	4	atención	1	302
Oper1	dar	3	espalda	1	302
Manif	anunciar	1	adjudicación	1	302
Oper2	recibir	1	ayuda	1	300
Real1	utilizar	1	método	1	299
FWC	ver	1	cara	1	297
ERROR	saber		no		297
FWC	reunir	1	requisito	1	297
FWC	repetir	1	vez	1	295
CausPlusFunc1	promover	1	desarrollo	1	295
Caus2Func1	hacer	2	ilusión	2	295
FWC	dar	4	mano	1	295
FWC	buscar	2	forma	1	294
Oper1	tener	3	certeza	2	293
FWC	quedar	1	remedio	1	293
Oper1	hacer	15	seguimiento	2	293
Oper2	recibir	1	apoyo	1	292
Oper2	escuchar	1	voz	2	292
CausFunc1	dar	4	crédito	1	292
Oper1	realizar	6	labor	1	290
Manif	marcar	3	diferencia	4	290
CausPerfFunc0	garantizar	3	derecho	1	290
Oper1	padecer	1	enfermedad	1	289
CausFunc0	hacer	2	programa	1	289
ContOper1	guardar	1	silencio	1	289
FWC	proteger	1	derecho	1	288
Oper1	tener	1	plan	1	287
Oper1	tener	1	opinión	3	287
CausFunc0	rendir	2	cuenta	N/A	287
Oper1	hacer	15	llamamiento	N/A	286
CausFunc1	dar	9	lección	1	286

CausFunc0	*causar*	1	*problema*	7	286
FWC	*abrir*	1	*boca*	1	286
FWC	*tener*	1	*niño*	1	285
Real1	*resolver*	2	*conflicto*	1	285
AntiReal3	*violar*	1	*derecho*	1	284
Caus2Func1	*dar*	9	*gusto*	3	284
CausFunc0	*abrir*	5	*posibilidad*	2	284
Oper1	*tener*	3	*motivo*	1	283
Oper1	*tener*	1	*dato*	1	283
CausFunc0	*sentar*	2	*base*	2	283
Caus2Func1	*dar*	9	*imagen*	1	283
CausFunc0	*aprobar*	1	*reglamento*	1	283
Caus2Func1	*dar*	9	*cabida*	1	282
FWC	*hacer*	2	*vez*	1	281
CausFunc1	*dar*	9	*derecho*	1	281
Oper1	*tener*	2	*imagen*	1	280
LiquFunc0	*quitar*	1	*vida*	2	280
CausFunc1	*poner*	4	*cara*	3	280
Oper1	*hacer*	15	*viaje*	1	279
Real1	*utilizar*	1	*técnica*	1	277
CausFunc0	*hacer*	2	*historia*	9	277
Oper1	*tener*	1	*edad*	2	276
CausFunc0	*establecer*	2	*norma*	3	276
CausFunc0	*encontrar*	6	*camino*	6	275
Real1	*ver*	1	*imagen*	1	274
Real1	*aprovechar*	1	*oportunidad*	1	274
Oper1	*tener*	2	*calidad*	1	273
Oper1	*hacer*	15	*llamada*	3	273
CausFunc0	*firmar*	4	*contrato*	1	273
FWC	*tener*	1	*libro*	1	272
Oper1	*cometer*	1	*delito*	1	272
Oper1	*tener*	1	*competencia*	2	271
CausFunc0	*desarrollar*	2	*sistema*	3	271
Oper1	*tener*	1	*condición*	2	270
ERROR	*tener*		*inconveniente*		269

Oper1	tener	1	estructura	3	269
Oper2	recibir	1	llamada	3	269
CausFunc1	poner	4	pie	5	269
Func1	existir	1	problema	7	269
FWC	dedicar	1	parte	1	269
Oper1	dar	3	cumplimiento	1	269
ContOper1	mantener	2	nivel	1	268
Oper1	tener	2	gusto	2	267
CausFunc0	hacer	2	amigo	1	267
CausFunc1	dar	9	instrucción	1	267
FWC	dar	4	dinero	1	267
ERROR	tener		porqué		266
FWC	tener	1	cuerpo	1	265
Oper1	realizar	6	proyecto	2	265
CausFunc0	producir	4	resultado	2	265
MinusReal1	gastar	1	dinero	1	265
Oper1	dar	3	paseo	1	265
Oper1	tener	1	cara	3	264
CausFunc1	establecer	2	contacto	1	264
Real1	lograr	2	objetivo	6	263
CausFunc0	desarrollar	2	proyecto	2	263
Real1	utilizar	1	término	2	262
FWC	meter	2	mano	1	262
CausFunc0	escribir	1	historia	3	262
Oper1	dar	3	voz	2	262
CausFunc1	dar	9	muerte	1	262
Oper2	recibir	1	atención	1	261
Manif	presentar	2	proyecto	2	261
Oper2	obtener	1	respuesta	2	261
Oper1	hacer	15	énfasis	2	261
Func1	existir	1	riesgo	1	261
FWC	tener	1	día	1	260
Oper1	realizar	6	operación	1	260
Oper1	tener	3	sueño	6	259
CausFunc1	dar	9	condición	5	259

CausFunc1	*dar*	9	*circunstancia*	1	259
CausFunc0	*crear*	1	*sistema*	3	259
Oper1	*realizar*	6	*acto*	2	258
AntiPermOper1	*prohibir*	1	*reproducción*	1	258
Oper1	*hacer*	15	*distinción*	2	258
Oper1	*añadir*	2	*comentario*	1	258
IncepOper1	*tomar*	4	*partido*	N/A	257
CausPlusFunc1	*elevar*	2	*nivel*	1	257
Oper1	*tener*	2	*memoria*	2	256
Oper2	*tener*	1	*honor*	2	256
Oper1	*tener*	2	*costumbre*	2	256
CausFunc0	*elaborar*	1	*plan*	1	256
CausFunc0	*abrir*	2	*cuenta*	6	256
ContOper1	*seguir*	11	*línea*	1	255
FWC	*poner*	1	*cosa*	3	255
Real1	*cumplir*	3	*condición*	5	255
IncepOper1	*tomar*	6	*posición*	11	254
Oper1	*tener*	1	*autoridad*	3	254
Oper1	*tener*	1	*privilegio*	1	253
FWC	*tener*	1	*constancia*	1	253
FWC	*referir*	1	*apartado*	1	253
Manif	*plantear*	1	*necesidad*	1	253
PermOper1	*permitir*	1	*desarrollo*	1	253
CausFunc1	*dar*	9	*conferencia*	1	253
CausFunc1	*crear*	1	*espacio*	2	253
Real1	*ver*	1	*foto*	1	252
Real1	*conseguir*	1	*resultado*	2	252
FWC	*ver*	1	*resultado*	2	250
Oper1	*realizar*	6	*obra*	1	250
ERROR	*hasta ahora*		*no*		250
IncepReal1	*abordar*	2	*problema*	7	250
FWC	*vivir*	1	*españa*	N/A	249
Real1	*ver*	1	*ejemplo*	1	249
Real1	*utilizar*	1	*tecnología*	2	249
Real1	*utilizar*	1	*palabra*	1	249

Oper1	*tener*	1	*color*	1	249
Oper2	*recibir*	1	*respuesta*	2	249
CausFunc1	*poner*	4	*acento*	2	249
CausFunc0	*establecer*	2	*mecanismo*	3	249
Oper1	*tener*	1	*título*	2	248
FinOper1	*perder*	1	*peso*	1	248
FinOper1	*perder*	1	*parte*	1	248
Real1	*dar*	N/A	*cara*	1	248
FWC	*solicitar*	2	*información*	1	247
ERROR	*escribir*		*no*		247
CausFunc1	*dar*	9	*soporte*	1	247
FWC	*tener*	1	*cabeza*	1	245
Copul	*ser*	1	*parte*	1	245
FWC	*decir*	1	*Santo*	N/A	245
Oper1	*ocupar*	3	*puesto*	2	244
Oper1	*tener*	1	*participación*	3	243
Oper2	*obtener*	1	*premio*	2	243
Caus2Func1	*dar*	9	*vergüenza*	1	243
FWC	*conocer*	1	*gente*	1	243
Oper1	*tener*	3	*sentimiento*	2	242
FWC	*levantar*	2	*cabeza*	1	242
CausFunc0	*hacer*	2	*plan*	1	242
ERROR	*demostrar*		*ser*		242
IncepOper1	*tomar*	6	*poder*	1	241
Copul	*ser*	1	*mujer*	1	241
Copul	*ser*	1	*estudiante*	1	241
Oper1	*realizar*	6	*cambio*	3	241
CausFunc1	*dejar*	4	*lugar*	4	241
CausFunc1	*dar*	9	*trabajo*	1	241
ERROR	*»*		*La*		240
FWC	*tener*	2	*cantidad*	4	240
FWC	*reunir*	1	*condición*	5	240
ManifFunc0	*plantear*	1	*pregunta*	1	240
Real1	*aplicar*	1	*ley*	3	240
FWC	*tener*	1	*casa*	1	239

CausFunc1	*hacer*	2	*camino*	6	239
Oper1	*ejercer*	2	*función*	1	239
Real1	*atender*	5	*necesidad*	1	239
IncepOper1	*tomar*	6	*palabra*	2	238
Oper1	*tener*	1	*propósito*	3	238
FWC	*tener*	1	*familia*	1	238
CausFunc1	*dar*	9	*apoyo*	1	237
Oper1	*hacer*	2	*juego*	2	236
FWC	*cambiar*	1	*nombre*	1	236
Oper2	*recibir*	1	*formación*	3	235
Oper2	*recibir*	1	*educación*	3	235
CausFunc1	*poner*	4	*límite*	1	235
FWC	*enviar*	4	*carta*	1	235
Oper1	*tener*	2	*validez*	1	234
Oper1	*tener*	2	*fama*	1	234
Oper1	*pasar*	1	*hambre*	2	234
LiquFunc0	*evitar*	2	*problema*	7	234
ERROR	*agenciar*		*español*		234
CausFunc1	*dar*	9	*señal*	1	233
Oper1	*tener*	1	*explicación*	1	232
Real1	*hablar*	3	*lengua*	2	232
Oper1	*tener*	1	*dimensión*	1	231
CausFunc1	*destacar*	4	*importancia*	1	231
FWC	*defender*	3	*interés*	1	231
CausFunc0	*construir*	1	*sociedad*	1	231
Oper1	*tener*	2	*habilidad*	1	230
ERROR	*procurar*		*no*		230
CausFunc1	*dejar*	1	*paso*	4	230
IncepOper1	*adoptar*	2	*forma*	1	230
CausMinusFunc1	*reducir*	2	*número*	1	229
Oper1	*tener*	2	*facultad*	1	228
Oper1	*seguir*	11	*ejemplo*	2	228
CausFunc1	*ofrecer*	1	*visión*	2	228
Oper1	*hacer*	15	*entrega*	5	228
CausFunc1	*dar*	9	*visión*	2	228

Oper1	tener	2	virtud	4	227
ContOper1	seguir	11	modelo	2	227
CausFunc1	despertar	1	interés	1	227
FWC	dar	4	dato	1	227
Oper1	tener	2	uso	1	226
CausFunc1	dar	9	impulso	4	226
Caus2Func1	costar	2	trabajo	1	226
IncepOper1	adoptar	2	actitud	1	226
Oper1	tener	3	prisa	3	225
Oper1	tener	3	deseo	1	225
Oper2	recibir	1	carta	1	225
ERROR	hacer		bien		225
Real1	comer	1	carne	2	225
FWC	buscar	2	ayuda	1	225
CausPlusFunc1	aumentar	2	riesgo	1	225
ERROR	tener		mano		224
ERROR	intentar		no		224
Oper1	hacer	15	juicio	3	224
FWC	comprar	1	disco	1	224
Oper1	tener	2	tamaño	1	223
Oper1	hacer	15	gala\galo	N/A	223
CausFunc0	dictar	3	sentencia	3	223
CausFunc1	causar	1	muerte	1	223
Oper1	tener	3	enfermedad	1	222
Oper1	hacer	15	reflexión	5	222
Oper1	hacer	15	gesto	3	222
CausFunc0	dejar	4	espacio	2	222
ERROR	·		no		221
FWC	tener	1	gente	1	221
ERROR	tambiÃ		©		221
Oper1	ocupar	3	posición	11	221
CausFunc0	hacer	2	lista	1	221
CausFunc0	encontrar	6	manera	1	221
CausFunc1	dar	9	inicio	2	221
Oper1	tener	2	incidencia	N/A	220

Func0	*pasar*	2	*siglo*	1	220
CausFunc1	*dar*	9	*libertad*	1	220
FWC	*citar*	2	*fuente*	15	220
Oper1	*tener*	2	*contenido*	2	219
CausFunc1	*abrir*	5	*espacio*	2	219
FWC	*visitar*	2	*página*	1	218
Oper2	*tener*	1	*apoyo*	1	218
PermOper1	*permitir*	1	*uso*	1	218
FinOper1	*perder*	1	*control*	1	218
ContOper1	*mantener*	2	*equilibrio*	1	218
FWC	*ver*	1	*figura*	2	217
Real1	*utilizar*	1	*recurso*	2	217
Oper1	*cursar*	2	*estudio*	3	217
IncepOper1	*adquirir*	2	*importancia*	1	217
Oper1	*tener*	2	*corazón*	2	216
Copul	*parecer*	1	*idea*	5	216
CausFunc0	*impartir*	1	*clase*	2	216
Oper1	*ejercer*	2	*control*	1	216
Oper1	*dejar*	12	*comentario*	1	216
CausFunc1	*producir*	4	*daño*	3	215
Manif	*presentar*	2	*información*	1	215
FWC	*necesitar*	2	*tiempo*	1	215
Oper1	*tener*	2	*coste*	2	214
FWC	*subir*	1	*escalera*	1	214
CausPlusFunc0	*mejorar*	1	*situación*	2	214
Oper1	*hacer*	2	*investigación*	3	214
CausFunc0	*establecer*	2	*criterio*	2	214
FWC	*comprar*	1	*producto*	2	214
CausMinusFunc0	*bajar*	5	*precio*	2	214
Caus2Func1	*atraer*	3	*atención*	1	214
Oper1	*tener*	2	*vigencia*	1	213
FWC	*tener*	1	*proyecto*	2	213
Real1	*satisfacer*	1	*demanda*	5	213
Func0	*pasar*	1	*cosa*	5	213
Oper1	*dar*	3	*beso*	1	213

FWC	*crear*	1	*mundo*	2	213
FWC	*construir*	1	*mundo*	2	213
Oper1	*tener*	2	*voluntad*	2	212
Oper1	*tener*	2	*actitud*	1	212
FWC	*incluir*	4	*información*	1	212
Func0	*haber*	2	*tiempo*	1	212
CausPlusFunc0	*desarrollar*	2	*capacidad*	4	212
ERROR	*«*		*hombre*		211
Oper1	*tener*	2	*fortuna*	2	211
Oper1	*tener*	2	*conexión*	1	211
IncepFunc0	*llegar*	3	*caso*	2	211
Oper1	*tener*	2	*sed*	2	210
FWC	*tener*	1	*aplicación*	1	210
Oper2	*sufrir*	3	*consecuencia*	1	210
ERROR	*hablar*		*no*		210
FWC	*esperar*	1	*respuesta*	2	210
FWC	*considerar*	2	*posibilidad*	2	210
FWC	*tener*	1	*elemento*	1	209
Oper1	*tener*	2	*alma*	2	209
Manif	*mostrar*	2	*interés*	1	209
Real2	*merecer*	1	*atención*	1	209
CausFunc1	*dar*	9	*publicidad*	1	209
CausFunc1	*dar*	9	*permiso*	1	209
CausFunc1	*dar*	9	*carácter*	1	209
Real1	*recorrer*	1	*camino*	6	208
Oper1	*realizar*	6	*función*	1	208
Manif	*presentar*	2	*informe*	1	208
FinOper1	*perder*	1	*sentido*	1	208
Oper1	*hacer*	15	*papel*	1	208
CausPlusFunc0	*favorecer*	1	*desarrollo*	1	208
Oper1	*ejercer*	2	*presión*	2	208
ERROR	*dar*		*no*		208
Real1	*hablar*	3	*idioma*	1	207
CausFunc1	*crear*	1	*ambiente*	1	207
FWC	*tomar*	1	*tiempo*	1	206

Oper2	tomar	1	sol	N/A	206
FWC	tener	1	raíz	2	206
FWC	tener	1	padre	1	206
ContOper1	seguir	11	curso	4	206
FWC	lavar	1	mano	1	206
FWC	decir	1	señor	5	206
Real1	utilizar	1	medio	3	205
Oper1	tener	2	grado	3	205
CausFunc1	establecer	2	base	1	205
Oper1	realizar	6	ejercicio	2	204
FWC	obtener	1	título	2	204
FWC	mandar	5	mensaje	2	204
CausFunc0	hacer	2	música	1	204
CausFunc0	sacar	4	foto	1	203
CausFunc0	elegir	1	presidente	1	203
CausFunc0	declarar	3	guerra	1	203
Real1	corregir	1	error	1	203
FWC	tener	1	novio	1	202
Oper1	realizar	6	seguimiento	2	202
Oper1	hacer	15	observación	5	202
FWC	conocer	1	verdad	3	202
FWC	conocer	1	historia	2	202
FWC	aprovechar	1	ocasión	2	202
ERROR	--		no		202
FWC	ver	1	artículo	3	201
CausFunc1	dar	9	problema	7	201
FWC	tener	1	futuro	1	200
Manif	presentar	2	resultado	2	200
CausFunc0	dictar	3	resolución	6	200
Oper1	dar	3	giro	6	200
CausFunc0	crear	1	grupo	1	200
FWC	contar	2	cosa	5	200
FWC	constituir	2	elemento	1	200
FWC	buscar	2	verdad	3	200
FWC	ver	1	tele	1	199

Real1	utilizar	1	programa	1	199
FWC	tener	1	gobierno	1	199
FWC	ser	1	ser	1	199
Oper2	recoger	7	información	2	199
Oper2	obtener	1	permiso	1	199
CausFunc0	impartir	1	curso	2	199
CausFunc0	dar	4	prueba	6	199
Func0	caber	1	posibilidad	2	199
FWC	analizar	2	situación	2	199
FWC	tener	1	serie	4	198
Oper1	tener	2	gracia	2	198
ERROR	querer		ser		198
CausFunc0	escribir	1	poema	1	198
FWC	transmitir	3	información	1	197
CausFunc0	surtir	1	efecto	1	197
ERROR	san		Josemaría		197
FWC	negar	2	existencia	1	197
FWC	facilitar	1	información	1	197
IncepOper1	cobrar	1	importancia	1	197
FWC	tener	1	lista	1	196
Oper1	tener	1	actividad	1	196
Oper2	sufrir	3	cambio	3	196
CausFunc0	provocar	2	reacción	3	196
CausFunc0	provocar	2	cambio	3	196
FinOper1	perder	1	oportunidad	1	196
CausFunc0	hacer	2	ley	3	196
Oper1	ejercer	2	poder	1	196
Copul	constituir	2	parte	1	196
Real1	conseguir	1	objetivo	6	196
FWC	buscar	2	empleo	2	196
FWC	tocar	16	tema	1	195
FWC	tener	1	agua	1	195
Oper2	sufrir	3	ataque	8	195
FWC	sacar	4	dinero	1	195
FWC	adoptar	1	decisión	2	195

IncepOper1	tomar	4	camino	6	194
FWC	tener	1	pelo	1	194
Oper2	sufrir	3	daño	3	194
FWC	pasar	2	vida	5	194
FWC	hacer	6	copia	1	194
CausFunc0	escribir	1	comentario	1	194
Caus2Func1	dar	9	asco	2	194
FWC	buscar	2	respuesta	2	194
Copul	ser	1	partidario	N/A	193
FWC	poner	1	punto	9	193
FWC	pasar	1	parte	1	193
FWC	pasar	1	fin	2	193
CausFunc0	hacer	2	fuerza	2	193
FWC	cruzar	1	frontera	1	193
FWC	ser	1	chico	1	192
Oper1	hacer	15	acto	2	192
FWC	aportar	2	dato	1	192
FWC	ver	1	página	1	191
Real1	utilizar	1	servicio	1	191
FWC	narrar	1	historia	3	191
FWC	lograr	2	resultado	2	191
CausFunc1	hacer	2	honor	1	191
FWC	clic	N/A	programa	4	191
FWC	buscar	2	manera	1	191
Real1	usar	1	programa	1	190
Oper1	tener	1	tradición	1	190
Oper1	tener	1	posición	11	190
FWC	tener	1	momento	1	190
Oper1	tener	2	existencia	1	190
Oper1	tener	3	dolor	3	190
Oper1	sentir	6	necesidad	1	190
CausFunc1	ofrecer	1	solución	1	190
FWC	levantar	2	mano	1	190
Oper1	ejercer	2	derecho	1	190
Oper2	recibir	1	orden	2	189

CausFunc0	*hacer*	2	*diferencia*	4	189
Func0	*existir*	1	*razón*	2	189
CausFunc0	*establecer*	2	*condición*	5	189
FWC	*ver*	1	*vida*	1	188
Real1	*usar*	1	*término*	2	188
ContOper1	*seguir*	11	*orden*	2	188
ERROR	*querer*		*no*		188
FWC	*quedar*	1	*tiempo*	1	188
Func0	*pasar*	2	*domingo*	1	188
FWC	*ir*	8	*cosa*	5	188
Real1	*ganar*	2	*guerra*	1	188
CausFunc0	*firmar*	4	*convenio*	1	188
Oper1	*tener*	1	*ejemplo*	1	187
Oper1	*presentar*	1	*dificultad*	3	187
IncepReal1	*abordar*	2	*cuestión*	2	187
FWC	*vivir*	2	*momento*	1	186
Oper1	*tener*	2	*relevancia*	1	186
FWC	*tener*	1	*pregunta*	1	186
CausFunc0	*proporcionar*	1	*servicio*	1	186
ERROR	*llamar*		«		186
CausFunc0	*introducir*	3	*cambio*	3	186
FWC	*tener*	1	*misión*	2	185
FWC	*tener*	1	*implicación*	2	185
Oper1	*realizar*	6	*evaluación*	3	185
Oper1	*meter*	N/A	*pata*	N/A	185
CausFunc0	*desarrollar*	2	*habilidad*	1	185
IncepOper1	*conseguir*	1	*trabajo*	6	185
FWC	*tener*	1	*persona*	1	184
Oper1	*tener*	2	*desarrollo*	1	184
FWC	*solicitar*	2	*ayuda*	1	184
Manif	*presentar*	2	*solicitud*	1	184
Func0	*pasar*	2	*sábado*	1	184
FWC	*obtener*	1	*dato*	1	184
FWC	*escuchar*	1	*palabra*	1	184
CausFunc1	*dar*	9	*curso*	2	184

CausFunc1	*prestar*	1	*ayuda*	1	183
CausFunc0	*hacer*	2	*revolución*	1	183
CausFunc0	*hacer*	2	*comparación*	1	183
Oper1	*tener*	1	*desperdicio*	N/A	182
Oper1	*presentar*	1	*característica*	3	182
Real1	*aplicar*	1	*medida*	3	182
FWC	*tener*	1	*componente*	1	181
CausMinusFunc0	*reducir*	2	*coste*	2	181
ERROR	*online*		*casino*		181
CausFunc0	*hacer*	2	*dinero*	1	181
Oper1	*hacer*	15	*cálculo*	1	181
ERROR	*explicar*		*porqué*		181
IncepOper1	*encontrar*	7	*lugar*	4	181
CausFunc0	*desarrollar*	2	*estrategia*	1	181
FWC	*compartir*	1	*experiencia*	1	181
Real1	*utilizar*	1	*lenguaje*	4	180
FWC	*tener*	1	*concepto*	1	180
FWC	*tener*	1	*arma*	1	180
Manif	*presentar*	2	*propuesta*	1	180
Manif	*plantear*	1	*posibilidad*	2	180
CausFunc0	*hacer*	2	*propuesta*	1	180
CausFunc1	*dar*	9	*salida*	1	180
IncepOper1	*asumir*	2	*papel*	1	180
Real1	*aplicar*	1	*política*	1	180
CausFunc0	*alcanzar*	1	*acuerdo*	3	180
Real1	*utilizar*	1	*herramienta*	2	179
Oper1	*tener*	1	*permiso*	1	179
Oper1	*tener*	1	*compromiso*	3	179
Oper1	*hacer*	15	*cuenta*	1	179
FWC	*dar*	4	*Silva*	N/A	179
FWC	*aprobar*	1	*plan*	1	179
FinFunc0	*transcurrir*	1	*plazo*	3	178
FWC	*tener*	1	*sitio*	1	178
FWC	*tener*	1	*equipo*	1	178
CausFunc1	*prestar*	1	*asistencia*	1	178

CausFunc0	*hacer*	2	*curso*	2	178
CausFunc0	*aportar*	2	*información*	1	178
ERROR	*san*		*Pablo*		177
ERROR	*pensar*		*no*		177
ERROR	*dar*		*Brown*		177
CausFunc0	*crear*	1	*red*	1	177
CausPlusFunc0	*ampliar*	1	*información*	1	177
PerfOper1	*alcanzar*	1	*grado*	3	177
FWC	*vivir*	2	*tiempo*	1	176
Oper1	*tener*	3	*sexo*	1	176
Oper1	*tener*	2	*sabor*	1	176
FWC	*tener*	1	*prueba*	6	176
FWC	*tener*	1	*grupo*	1	176
FWC	*tener*	1	*año*	2	176
Func0	*pasar*	2	*viernes*	1	176
CausFunc0	*establecer*	2	*límite*	1	176
FWC	*defender*	3	*libertad*	1	176
CausFunc1	*dar*	9	*seguridad*	3	176
CausFunc1	*dar*	9	*satisfacción*	1	176
Oper1	*vivir*	2	*experiencia*	1	175
Manif	*pronunciar*	1	*palabra*	2	175
CausFunc0	*hacer*	2	*pausa*	5	175
Oper1	*ejecutar*	2	*programa*	4	175
FWC	*contener*	4	*sustancia*	1	175
FWC	*buscar*	2	*vida*	1	175
ERROR	"		*La*		174
Real1	*usar*	1	*sistema*	3	174
Oper1	*tener*	2	*curiosidad*	1	174
FWC	*resumir*	1	*cuenta*	1	174
FWC	*realizar*	6	*serie*	4	174
CausFunc0	*provocar*	2	*muerte*	1	174
CausFunc0	*establecer*	2	*diferencia*	4	174
Real1	*contestar*	2	*pregunta*	1	174
Real1	*cantar*	1	*canción*	1	174
FWC	*vender*	1	*producto*	2	173

PerfOper1	*tomar*	4	*precaución*	2	173
FWC	*tener*	1	*fundamento*	1	173
Func1	*quedar*	1	*duda*	2	173
FWC	*decir*	1	*pepito*	1	173
CausFunc1	*dar*	4	*rienda*	2	173
FWC	*actualizar*	1	*entrada*	9	173
ERROR	*vivir*		*no*		172
Oper1	*tener*	1	*servicio*	1	172
Oper1	*tener*	2	*formación*	3	172
Oper1	*tener*	2	*finalidad*	3	172
CausFunc0	*provocar*	2	*daño*	3	172
ContOper1	*mantener*	2	*control*	1	172
Oper1	*celebrar*	2	*reunión*	1	172
ERROR	*volver*		*loco*		171
ERROR	*resultar*		*no*		171
FWC	*recoger*	7	*dato*	1	171
CausFunc0	*producir*	4	*aumento*	2	171
FWC	*pedir*	2	*dinero*	1	171
CausFunc0	*establecer*	2	*marco*	1	171
FWC	*comprar*	1	*casa*	1	171
FWC	*cambiar*	1	*forma*	1	171
CausPlusFunc0	*ampliar*	1	*dato*	1	171
PerfOper1	*tomar*	1	*cuerpo*	1	170
Oper1	*tener*	2	*respeto*	2	170
Oper1	*tener*	1	*referencia*	2	170
ERROR	*resultar*		*ser*		170
CausFunc0	*garantizar*	3	*acceso*	3	170
CausFunc0	*establecer*	2	*conexión*	1	170
ERROR	*entender*		*no*		170
CausFunc0	*brindar*	N/A	*oportunidad*	1	170
Real1	*alcanzar*	1	*meta*	1	170
IncepOper1	*tomar*	4	*contacto*	1	169
FWC	*tener*	1	*cáncer*	1	169
FWC	*requerir*	1	*esfuerzo*	2	169
Oper1	*realizar*	6	*viaje*	1	169

FWC	*ganar*	5	*tiempo*	1	169
CausFunc0	*escribir*	1	*texto*	1	169
Real1	*enfrentar*	1	*problema*	7	169
FWC	*decir*	1	*adiós*	1	169
FWC	*dar*	4	*título*	2	169
ContOper1	*seguir*	11	*proceso*	3	168
Caus2Func1	*imponer*	1	*pena*	12	168
FWC	*hacer*	2	*milagro*	1	168
CausFunc0	*hacer*	2	*contacto*	1	168
CausFunc0	*establecer*	2	*ley*	3	168
Real1	*contar*	2	*cuento*	1	168
FWC	*tener*	1	*mujer*	1	167
Oper1	*hacer*	15	*lectura*	2	167
FWC	*hacer*	2	*gobierno*	1	167
CausFunc0	*escribir*	1	*poesía*	1	167
FWC	*decir*	1	*Jesús*	1	167
IncepOper1	*asumir*	5	*riesgo*	1	167